Selected Asymptotic Methods with Applications to Electromagnetics and Antennas

Synthesis Lectures on Computational Electromagnetics

Editor
Constantine A. Balanis, *Arizona State University*

Synthesis Lectures on Computational Electromagnetics will publish 50- to 100-page publications on topics that include advanced and state-of-the-art methods for modeling complex and practical electromagnetic boundary value problems. Each lecture develops, in a unified manner, the method based on Maxwell's equations along with the boundary conditions and other auxiliary relations, extends underlying concepts needed for sequential material, and progresses to more advanced techniques and modeling. Computer software, when appropriate and available, is included for computation, visualization and design. The authors selected to write the lectures are leading experts on the subject that have extensive background in the theory, numerical techniques, modeling, computations and software development.

The series is designed to:

- Develop computational methods to solve complex and practical electromagnetic boundary-value problems of the 21st century.

- Meet the demands of a new era in information delivery for engineers, scientists, technologists and engineering managers in the fields of wireless communication, radiation, propagation, communication, navigation, radar, RF systems, remote sensing, and biotechnology who require a better understanding and application of the analytical, numerical and computational methods for electromagnetics.

Selected Asymptotic Methods with Applications to Electromagnetics and Antennas
George Fikioris, Ioannis Tastsoglou, and Odysseas N. Bakas
2013

Double-Grid Finite-Difference Frequency-Domain (DG-FDFD) Method for Scattering from Chiral Objects
Erdogan Alkan, Veysel Demir, Atef Elsherbeni, and Ercument Arvas
2013

Multiresolution Frequency Domain Technique for Electromagnetics
Mesut Gokten, Atef Elsherbeni, and Ercument Arvas
2012

Scattering Analysis of Periodic Structures Using Finite-Difference Time-Domain Method
Khaled ElMahgoub, Fan Yang, and Atef Elsherbeni
2012

Introduction to the Finite-Difference Time-Domain (FDTD) Method for Electromagnetics
Stephen D. Gedney
2011

Analysis and Design of Substrate Integrated Waveguide Using Efficient 2D Hybrid Method
Xuan Hui Wu and Ahmed A. Kishk
2010

An Introduction to the Locally-Corrected Nyström Method
Andrew F. Peterson and Malcolm M. Bibby
2009

Transient Signals on Transmission Lines: An Introduction to Non-Ideal Effects and Signal Integrity Issues in Electrical Systems
Andrew F. Peterson and Gregory D. Durgin
2008

Reduction of a Ship's Magnetic Field Signatures
John J. Holmes
2008

Integral Equation Methods for Electromagnetic and Elastic Waves
Weng Cho Chew, Mei Song Tong, and Bin Hu
2008

Modern EMC Analysis Techniques Volume II: Models and Applications
Nikolaos V. Kantartzis and Theodoros D. Tsiboukis
2008

Modern EMC Analysis Techniques Volume I: Time-Domain Computational Schemes
Nikolaos V. Kantartzis and Theodoros D. Tsiboukis
2008

Particle Swarm Optimization: A Physics-Based Approach
Said M. Mikki and Ahmed A. Kishk
2008

Three-Dimensional Integration and Modeling: A Revolution in RF and Wireless Packaging
Jong-Hoon Lee and Manos M. Tentzeris
2007

Electromagnetic Scattering Using the Iterative Multiregion Technique
Mohamed H. Al Sharkawy, Veysel Demir, and Atef Z. Elsherbeni
2007

Electromagnetics and Antenna Optimization Using Taguchi's Method
Wei-Chung Weng, Fan Yang, and Atef Elsherbeni
2007

Fundamentals of Electromagnetics 1: Internal Behavior of Lumped Elements
David Voltmer
2007

Fundamentals of Electromagnetics 2: Quasistatics and Waves
David Voltmer
2007

Modeling a Ship's Ferromagnetic Signatures
John J. Holmes
2007

Mellin-Transform Method for Integral Evaluation: Introduction and Applications to Electromagnetics
George Fikioris
2007

Perfectly Matched Layer (PML) for Computational Electromagnetics
Jean-Pierre Bérenger
2007

Adaptive Mesh Refinement for Time-Domain Numerical Electromagnetics
Costas D. Sarris
2006

Frequency Domain Hybrid Finite Element Methods for Electromagnetics
John L. Volakis, Kubilay Sertel, and Brian C. Usner
2006

Exploitation of A Ship's Magnetic Field Signatures
John J. Holmes
2006

Support Vector Machines for Antenna Array Processing and Electromagnetics
Manel Martínez-Ramón and Christos Christodoulou
2006

The Transmission-Line Modeling (TLM) Method in Electromagnetics
Christos Christopoulos
2006

Computational Electronics
Dragica Vasileska and Stephen M. Goodnick
2006

Higher Order FDTD Schemes for Waveguide and Antenna Structures
Nikolaos V. Kantartzis and Theodoros D. Tsiboukis
2006

Introduction to the Finite Element Method in Electromagnetics
Anastasis C. Polycarpou
2006

MRTD(Multi Resolution Time Domain) Method in Electromagnetics
Nathan Bushyager and Manos M. Tentzeris
2006

Mapped Vector Basis Functions for Electromagnetic Integral Equations
Andrew F. Peterson
2006

Selected Asymptotic Methods with Applications to Electromagnetics and Antennas
George Fikioris, Ioannis Tastsoglou, and Odysseas N. Bakas

ISBN: 978-3-031-00588-6   paperback
ISBN: 978-3-031-01716-2   ebook

DOI 10.1007/978-3-031-01716-2

A Publication in the Springer series
SYNTHESIS LECTURES ON COMPUTATIONAL ELECTROMAGNETICS

Lecture #31
Series Editor: Constantine A. Balanis, *Arizona State University*
Series ISSN
Synthesis Lectures on Computational Electromagnetics
Print 1932-1252   Electronic 1932-1716

Selected Asymptotic Methods with Applications to Electromagnetics and Antennas

George Fikioris, Ioannis Tastsoglou, and Odysseas N. Bakas
School of Electrical and Computer Engineering
National Technical University of Athens
Zografou, Athens, Greece

SYNTHESIS LECTURES ON COMPUTATIONAL ELECTROMAGNETICS
#31

ABSTRACT

This book describes and illustrates the application of several asymptotic methods that have proved useful in the authors' research in electromagnetics and antennas. We first define asymptotic approximations and expansions and explain these concepts in detail. We then develop certain prerequisites from complex analysis such as power series, multivalued functions (including the concepts of branch points and branch cuts), and the all-important gamma function. Of particular importance is the idea of analytic continuation (of functions of a single complex variable); our discussions here include some recent, direct applications to antennas and computational electromagnetics.

Then, specific methods are discussed. These include integration by parts and the Riemann-Lebesgue lemma, the use of contour integration in conjunction with other methods, techniques related to Laplace's method and Watson's lemma, the asymptotic behavior of certain Fourier sine and cosine transforms, and the Poisson summation formula (including its version for finite sums). Often underutilized in the literature are asymptotic techniques based on the Mellin transform; our treatment of this subject complements the techniques presented in our recent Synthesis Lecture on the exact (not asymptotic) evaluation of integrals.

Throughout, we provide illustrative examples. Some of them are applications to special functions of mathematical physics. Others, taken from our published research, include the application of elementary methods to develop certain simple formulas for transmission lines, examples illustrating the difficulties in solving fundamental integral equations of antenna theory, an examination of the fundamentals of the Method of Auxiliary Sources (MAS), and a study of the near fields of certain unusual types of radiators.

KEYWORDS

asymptotics, antenna theory, wire antennas, electromagnetic theory, complex variables, special functions

In memory of John G. Fikioris (1931–2012)

Contents

Preface xvii

    References xviii

1   Introduction: Simple Asymptotic Approximations 1

  1.1 Far Field of Linear Antenna 1

  1.2 Period of Simple Pendulum: Small Oscillations 4

  1.3 A Differential Equation 7

      1.3.1 "Series Solution" 8

      1.3.2 The Solution as an Integral: Integration by Parts 10

  1.4 Asymptotic Approximations for High-SWR Transmission Lines 14

      1.4.1 Exact Formulas 14

      1.4.2 Further Exact Formulas: Large- and Small- g regions 15

      1.4.3 Asymptotic Formulas for the Large-g Region 16

      1.4.4 Asymptotic Formulas for the Small-g Region 18

  1.5 Supplementary Remarks and Further Reading 18

  1.6 Problems 19

      References 20

2   Asymptotic Approximations Defined 23

  2.1 Definitions 23

  2.2 Remarks and Examples 24

  2.3 Compound Asymptotic Approximations 27

      2.3.1 Elementary Example 27

      2.3.2 Bessel Function of Order Zero 28

  2.4 Asymptotic Expansions 30

  2.5 Historical and Supplementary Remarks 31

  2.6 Problems 32

      References 33

3   Concepts from Complex Variables 35

  3.1 Gamma Function and Related Functions 35

3.2   Power Series 39

3.3   Analytic Continuation 41

    3.3.1   Removable Singularities 41

    3.3.2   Geometric Series 41

    3.3.3   Analytic Continuation Defined: Uniqueness 41

    3.3.4   Further Examples 43

    3.3.5   Integrals that are Analytic Functions of a Parameter 43

3.4   Multivalued Functions and Branch Points 44

    3.4.1   Square Root 44

    3.4.2   Further Examples 46

    3.4.3   The Point at Infinity 47

3.5   Branches and Principal Values of Multivalued Functions 48

    3.5.1   Square Root 48

    3.5.2   Logarithm and Powers Other Than the Square Root 50

    3.5.3   The function $\sqrt{z^2 - 1}$ 51

    3.5.4   Values on Branch Cuts 52

3.6   Applications to Antennas and Electromagnetics: Nonsolvability 53

    3.6.1   Hallén's and Pocklington's Equations with the Approximate Kernel . . . 53

    3.6.2   Integral Equation Related to the Method of Auxiliary Sources (MAS) . 55

3.7   Supplementary Remarks and Further Reading 59

3.8   Problems 60

    References 66

4   Laplace's Method and Watson's Lemma **69**

4.1   Laplace's Method 69

    4.1.1   Simple Example 69

    4.1.2   Related Examples 71

    4.1.3   Stirling's Formula: Leading Term 73

    4.1.4   An Application to the Thin-Wire Circular-Loop Antenna 74

4.2   Watson's Lemma 77

    4.2.1   Statement of Lemma and Motivation 77

    4.2.2   Remarks and Extensions 78

    4.2.3   Examples 79

    4.2.4   Stirling's Formula Revisited and Lagrange Inversion Theorem 80

    4.2.5   An Application to the Method of Auxiliary Sources 82

4.3   Additional Remarks 84

4.4   Problems 85

References 86

5 Integration by Parts and Asymptotics of Some Fourier Transforms **89**

5.1 Integration by Parts and Laplace Transforms 89
    5.1.1 Complementary Error Function 89
    5.1.2 Remarks 91
5.2 Integration by Parts and Fourier Transforms 92
    5.2.1 Simple Example: Riemann-Lebesgue Lemma 92
    5.2.2 Remarks on the Lemma 93
    5.2.3 Simple Example Continued 93
    5.2.4 Example with Zero Boundary Terms 94
5.3 More on Fourier Transforms 96
5.4 Applications to Wire Antennas 97
    5.4.1 On the Kernels of Hallén's and Pocklington's Equations 98
    5.4.2 Behavior of Current Near Delta-Function Generator 100
5.5 Problems 101
    References 103

6 Poisson Summation Formula and Applications **105**

6.1 Doubly Infinite Sums 105
    6.1.1 Formula and its Derivation 105
    6.1.2 Remarks 106
    6.1.3 A First Example 107
    6.1.4 Application: Infinite Linear Array of Traveling-Wave Currents 110
    6.1.5 Application: Coupled Pseudopotential Arrays 114
6.2 Finite Sums 115
    6.2.1 Formula and Proof 115
    6.2.2 Remarks 117
    6.2.3 Elementary Example 118
    6.2.4 Continuous Functions with Equal Endpoint Values 118
    6.2.5 Application: Cylindrical Array of Traveling-Wave Currents 119
6.3 Problems 122
    References 126

7 Mellin-Transform Method for Asymptotic Evaluation of Integrals **129**

7.1 Summary of Mellin-Transform Method 129
7.2 Lemmas for Residue Calculations 132

7.3	Simple Example	133
7.4	On the Convergence of Mellin-Barnes Integrals	135
7.5	Application to Highly Directive Current Distributions	136
7.6	Further Reading	138
7.7	Problems	138
	References	140

8   More Applications to Wire Antennas **143**

8.1	Problem Pertaining to Magnetic Frill Generator	143
	8.1.1 Statement of Problem	143
	8.1.2 Preliminaries	144
	8.1.3 Derivation of Eq. 8.6	145
8.2	Oscillations with the Approximate Kernel: Case of Delta-Function Generator	146
	8.2.1 Integral Equation: Nonsolvability	146
	8.2.2 Numerical Method: Solution for Nonzero Discretization Length	147
	8.2.3 Asymptotic Approximation for Small Discretization Length	149
8.3	On the Near Field Due to Oscillating Current	152
	8.3.1 Statement of Problem	152
	8.3.2 Derivation of Eq. 8.46	153
8.4	Supplementary Remarks	154
8.5	Problems	155
	References	155

A   Special Functions **159**

A.1	Preliminaries	159
A.2	Exponential, Sine, and Cosine Integrals	160
	A.2.1 Definitions and Small-Argument Expansions	160
	A.2.2 Large-Argument Expansions	161
A.3	Complete Elliptic Integral of the First Kind	162
A.4	Bessel and Hankel Functions	163
	A.4.1 Definitions and Small-Argument Asymptotic Approximations	163
	A.4.2 Large-Argument Asymptotic Expansions	164
	A.4.3 Large-Order Asymptotic Approximations	166
	A.4.4 Addition Theorem for Hankel Function of Order Zero	166
A.5	Modified Bessel Functions	167

A.6   Generalized Hypergeometric Functions 167

A.7   Problems 168

        References 171

B   On the Convergence/Divergence of Definite Integrals 173

B.1   Some Remarks on Our Rules 173

B.2   Rules for Determining Convergence/Divergence 174

B.3   Examples 176

        References 177

Authors' Biographies 179

Index 181

Preface

Ever since the 1973 publication of the now-classic *Radiation and Scattering of Waves* by L. B. Felsen and N. Marcuvitz [1], asymptotic methods related to saddle points—such as the methods of steepest descents and stationary phase—have been well-covered in the electromagnetics literature. The purpose of this book is to present a number of other methods that, over the years, have been shown to be very useful in the authors' research in electromagnetics and antennas. While most of these methods deal with the asymptotic expansion of integrals, many are unrelated, or only marginally related, to saddle-point methods and their variants; some are much easier to understand.

The motto of R. W. Hamming's 1962 book on numerical methods was "The purpose of computing is insight, not numbers" [2]. Today, when it is much easier to obtain numbers using computer codes, we should view asymptotics in the way Hamming viewed computing. While their importance for producing precise numbers has lessened, asymptotics—when applicable—can greatly aid in providing insight to physical problems. The authors of a recent article in the *Notices of the American Mathematical Society* iterate this point in the slightly different context of closed-form solutions. In their conclusion, they note, "we feel strongly that the value of closed forms increases as the complexity of the objects we manipulate computationally and inspect mathematically grows" [3]. We illustrate throughout this book that asymptotics are particularly important for obtaining insight into certain fundamental problems of electromagnetics and antennas.

This work grew out of existing notes for a graduate course called "Methods of Applied Mathematics for Electromagnetic Fields" taught by George Fikioris for many years at the School of Electrical and Computer Engineering, National Technical University of Athens, Greece. Much of Chapter 7 came from a seminar written by G. Fikioris and Professor Dionisios Margetis and presented in 1996 at the Air Force Research Laboratory, Sensors Directorate, where G. Fikioris was then working.

Apart from an undergraduate-level exposition to electromagnetics, antennas, real analysis, and elementary complex analysis, this work is self-contained. In Chapters 1 and 2, asymptotic approximations and expansions are defined and explained in detail via elementary examples. We have attempted to provide clear explanations of fundamental concepts from which the methods evolve and to emphasize ideas that students find confusing; accordingly, in Chapter 3 we discuss prerequisites from complex analysis such as power series, multivalued functions (including the concepts of branch points and branch cuts), and the all-important gamma function, which is used in many places throughout this book. Of particular importance is the idea of analytic continuation (of functions of a single complex variable); our exposition here includes some recent, direct applications to computational electromagnetics and antennas. Chapters 4–8 discuss specific asymptotic methods. These include Laplace's method and Watson's lemma (Chapter 4), integration by parts, the Riemann-Lebesgue lemma, the asymptotic behavior of certain Fourier sine and cosine integrals (Chapter 5), and the powerful Poisson's summation formula, including its version for finite sums (Chapter 6). Often underutilized in the literature are asymptotic techniques based on the Mellin transform; our treatment of this subject (Chapter 7) complements the techniques presented in our recent Synthesis Lecture on the *exact* (not asymptotic) evaluation of integrals [4]. Applications related to computational techniques for wire antennas that combine previously developed asymptotic methods are discussed in Chapter 8. Essential to our development is Appendix A: it discusses, in detail, properties of special functions (such as Bessel functions) used in this book.

Throughout, we provide illustrative examples. Some are applications to special functions of mathematical physics. Others, taken from our published research, include the application of elementary methods to develop certain simple formulas for transmission lines, examples illustrating the difficulties in solving fundamental integral equations of antenna theory, an examination of the fundamentals of the Method of Auxiliary Sources (MAS), and a study of the near fields of certain unusual types of radiators.

We emphasize the understanding and application of techniques rather than the underlying proofs. To this end, some theorems are quoted, without proof, from the literature; in other cases, we provide nonrigorous derivations. Each chapter contains a number of exercises, many of which are essential for comprehension and practice. We provide a number of references for each topic and frequently give guides for further reading. We feel that the *Digital Library of Mathematical Functions* [5] (DLMF; we describe this work in Appendix A) will be a standard reference work for years to come; we have thus attempted to cite the DLMF whenever it is relevant.

Our mathematical notation is consistent with the DLMF. Thus, $\mathbb{Z}$, $\mathbb{Q}$, $\mathbb{R}$, and $\mathbb{C}$, respectively denote the sets of integer, rational, real, and complex numbers; and $\mathcal{R}z$, $\mathcal{I}z$, ph z, and $\bar{z}$ signify the real part, imaginary part, phase, and complex conjugate of z. In order to be consistent with related publications, both time dependences $e^{j\omega t}$ and $e^{-i\omega t}$ appear in our applications sections, where $\omega = kc$ is the angular frequency, k is the free-space wavenumber, and c is the speed of light.

Finally, the senior author is greatly indebted to Professor Tai Tsun Wu and the late Professor George F. Carrier for originally introducing him to asymptotics.

REFERENCES

[1] L. B. Felsen and N. Marcuvitz, *Radiation and Scattering of Waves*. Piscataway, NJ: IEEE Press, 1994 (Reprint of 1973 Ed.) DOI: 10.1109/9780470546307. xvii

[2] R. W. Hamming, *Numerical Methods for Scientists and Engineers*. New York: McGraw-Hill, 1962. xvii

[3] J. M. Borwein and R. E. Crandall, "Closed Forms: What They Are and Why We Care," *Notices of the AMS*, vol. 60, no. 1, pp. 50–65, January 2013. DOI: 10.1090/noti936. xvii

[4] G. Fikioris, *Mellin-transform method for integral evaluation: Introduction and applications to electromagnetics*. (Synthesis Lectures on Computational Electromagnetics #13). Morgan and Claypool Publishers, 2007. DOI: 10.2200/S00076ED1V01Y200612CEM013. xviii

[5] F. W. J. Olver, D. W. Lozier, R. F. Boisvert, and C. W. Clark, *Digital Library of Mathematical Functions*, National Institute of Standards and Technology from `http://dlmf.nist.gov/`. xviii

George Fikioris, Ioannis Tastsoglou, and Odysseas N. Bakas
July 2013

CHAPTER 1

Introduction: Simple Asymptotic Approximations

We can loosely describe an asymptotic approximation as being a solution to a problem whose accuracy improves as a problem parameter increases (or decreases, or approaches some finite value). We usually seek asymptotic approximations that are in closed form, or in some sense simple.

We will provide a precise definition of an asymptotic approximation in the next chapter. In the present chapter, we introduce some basic ideas of asymptotics by presenting effective asymptotic approximations for some undergraduate-level problems of mathematics, physics, and electrical engineering. The final asymptotic expressions come up rather easily and are evidently closed forms (we make no attempt to strictly define what a closed form is, see the relevant discussions in [1]). The present chapter thus demonstrates that we can obtain useful asymptotic approximations using elementary tools. Moreover, and similar to [1], this chapter also aims to stress that closed-form approximations—and closed-form asymptotic approximations in particular—may be preferable to complex exact solutions.

1.1   FAR FIELD OF LINEAR ANTENNA

Many transmitting antenna problems amount to finding the electromagnetic field generated by a given current density $\mathbf{J}e^{j\omega t}$ within a volume V. The usual first step is to calculate the vector potential $\mathbf{A}$ using the well-known equation [2]

$$\mathbf{A} = \frac{\mu}{4\pi} \iiint\limits_{V} \mathbf{J} \frac{e^{-jkR}}{R} dV , \tag{1.1}$$

where μ is the permeability of the surrounding medium, $k = \omega/c$ is the wavenumber, and $R = |\mathbf{R}| = |\mathbf{r}_p - \mathbf{r}'|$ is the distance from a source point within V to the observation point P, as shown in Fig. 1.1. A particularly simple case is that of a current filament of length $2h$—this is a model for the straight, thin wire antenna—assumed to lie along the z-axis, with the center at the origin. In this case, the triple integral in Eq. (1.1) reduces to a single integral and $\mathbf{A} = \hat{z}A_z$, with

$$A_z = \frac{\mu}{4\pi} \int_{-h}^{h} I\left(z'\right) \frac{e^{-jkR}}{R} dz' , \tag{1.2}$$

where $I\left(z'\right)$ is the current distribution along the antenna (in ampere) and

$$R = \sqrt{r^2 + \left(z'\right)^2 - 2rz'\cos\theta} \, , \tag{1.3}$$

see Fig. 1.2.

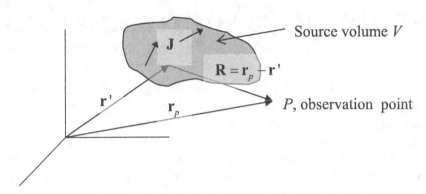

Figure 1.1: Vectors relevant to Eq. (1.1).

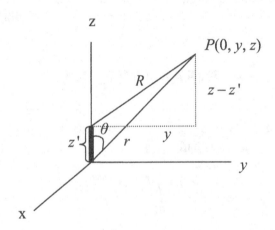

Figure 1.2: Distances relevant to linear antenna along the z-axis and Eqs. (1.2) and (1.3).

In most cases we are interested in the far field, where the observation point is far from the antenna in the sense that $kr \gg 1$ and $r \gg h$. Then $r \gg |z'|$ for all z' appearing in Eq. (1.2). We can thus approximate the denominator of Eq. (1.2) using $R \sim r$. In e^{-jkR}, however, we use the improved approximation

$$R \sim r - z'\cos\theta \, . \tag{1.4}$$

Eq. (1.4) is readily understood through Fig. 1.3: Subject to $r \gg h$, the distances R and r become nearly parallel and differ by the projection of one onto the other. Eq. (1.4) can also be shown using a Maclaurin expansion, see Problem 1.1. Our approximations amount to setting

$$\frac{e^{-jkR}}{R} \sim e^{-jkr} \frac{e^{jkz' \cos \theta}}{r} \tag{1.5}$$

in Eq. (1.2). The resulting equation has separate r- and θ-factors,

$$A_z \sim \frac{\mu}{4\pi} \frac{e^{-jkr}}{r} \int_{-h}^{h} I\left(z'\right) e^{jkz' \cos \theta} dz' \; . \tag{1.6}$$

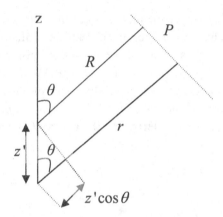

Figure 1.3: Parallel-ray approximation for far field calculations of a linear antenna.

By the nature of our derivation, the accuracy of the approximate equation Eq. (1.6) improves as r increases. Eq. (1.6) is simpler than the exact Eq. (1.2) and, for the usual case when one is interested in the far field, much more revealing and useful. For example, Eq. (1.6) immediately shows that, in the far field, $\mathbf{A}$ is an outward-traveling spherical wave, emanating from the origin. Given a closed-form current distribution $I\left(z'\right)$, Eq. (1.6) can lead to a closed-form solution for $\mathbf{A}$. For example, for the usual sinusoidal current distribution $I\left(z'\right) = I_m \sin k\left(h - |z'|\right)$, Eq. (1.6) gives

$$A_z \sim \frac{\mu I_m}{2\pi} \frac{e^{-jkr}}{kr} \frac{\cos\left(kh \cos \theta\right) - \cos kh}{\sin^2 \theta} \; . \tag{1.7}$$

Eq. (1.7) in turn leads to formulas for the electric and magnetic field, cf. [2] and [3]. Equations like (1.7) also hold for more accurate closed-form current distributions [3]. Finally, far-field equations like (1.6) and (1.7) form the starting point for defining and studying the array factor, an important concept for antenna arrays [2].

Asymptotic approximations like Eq. (1.7) are useful in physical problems both for numerical calculations and for obtaining understanding. The reader is likely to be familiar with other approximations in electromagnetics that one can consider asymptotic. The near electric field of an infinitesimal dipole, for example, is the dominant term, proportional to $1/r^3$, of the exact electric field [2]; Geometrical Optics (GO), as well as improvements to GO such as the Geometrical and Physical Theories of Diffraction, are asymptotic theories because their accuracy improves for high frequencies [4]; and circuit theory (i.e., Kirchhoff's laws together with the current-voltage relations) can be derived from Maxwell's equations for the case of low frequencies [4].

1.2   PERIOD OF SIMPLE PENDULUM: SMALL OSCILLATIONS

We now turn to an example from elementary physics that we will examine in some detail. The "simple pendulum" is an idealization of a number of real-life oscillators: a point mass (or pendulum bob) m is suspended by a massless rigid rod (or a massless, nonstrechable, always taut cord) of length ℓ, see Fig. 1.4. There is no friction or air resistance. We displace the bob from the vertical (equilibrium) position by an angle θ_m—assumed here for simplicity to satisfy $0 < \theta_m < \pi/2$— and, at time $t = 0$, leave it free. Under these conditions, periodic motion occurs within a circular arc, between the angles $\pm\theta_m$. We will determine the period T of oscillation.

Figure 1.4: Simple pendulum and forces acting on it. The restoring force $-mg\sin\theta(t)$ is the tangential component of the net force.

If $\theta(t)$ denotes the angular displacement from the vertical, then $\theta(0) = \theta_m$ and $\theta'(0) = 0$, where the prime denotes d/dt, with last condition meaning that the initial velocity is zero. The force perpendicular to the rod and along the tangential direction, called the "restoring force," is $-mg\sin[\theta(t)]$ where $g = 9.81$ m/sec^2 is the acceleration due to gravity, see Fig. 1.4. By Newton's second law of motion, this force equals the acceleration along the tangential direction, $ms''(t)$, where $s(t) = \ell\theta(t)$ is the circular arclength measured from the vertical position. It follows that

$$\theta''(t) + \frac{g}{\ell}\sin[\theta(t)] = 0 ,\qquad(1.8)$$

in which the mass m has dropped out. Eq. (1.8) is the "equation of motion." It is a nonlinear, second-order differential equation for $\theta(t)$ that is to be solved together with the initial conditions $\theta(0) = \theta_m$ and $\theta'(0) = 0$. An alternative form can be found by multiplying Eq. (1.8) by $\theta'(t)$ and recognizing that $\theta'(t)\theta''(t) = (1/2)d[\theta'(t)]^2/dt$ and $\theta'(t)\sin[\theta(t)] = -d\cos[\theta(t)]/dt$. Thus, $[\theta'(t)]^2 = \frac{2g}{\ell}\cos[\theta(t)] + C$. By the two initial conditions, the constant C is $C = -\frac{2g}{\ell}\cos\theta_m$. Therefore,

$$\theta'(t) = -\sqrt{\frac{2g}{\ell}[\cos\theta(t) - \cos\theta_m]},\qquad 0 < t < T/4 ,\qquad(1.9)$$

where the sign was chosen because the velocity is negative for $0 < t < T/4$. An equivalent expression is

$$\theta'(t) = -2\sqrt{\frac{g}{\ell}\left[\sin^2\frac{\theta_m}{2} - \sin^2\frac{\theta(t)}{2}\right]},\qquad 0 < t < T/4 .\qquad(1.10)$$

Each of Eqs. (1.9) and (1.10) are first-order, nonlinear differential equations for $\theta(t)$ that do not involve m, to be solved together with $\theta(0) = \theta_m$. Eq. (1.10) can also be derived directly (without appealing to Newton's law of motion or to Eq. (1.8)) from the principle of conservation of mechanical energy [5].

We can use Eq. (1.10) and its initial condition to obtain an explicit expression for the inverse $t(\theta)$ of $\theta(t)$. Since

$$\frac{dt}{d\theta} = -\frac{1}{2}\sqrt{\frac{\ell}{g}}\frac{1}{\sqrt{\sin^2(\theta_m/2) - \sin^2(\theta/2)}},\qquad 0 < \theta < \theta_m\qquad(1.11)$$

and since $t(\theta_m) = 0$, the desired explicit expression is

$$t(\theta) = \frac{1}{2}\sqrt{\frac{\ell}{g}}\int_\theta^{\theta_m}\frac{d\phi}{\sqrt{\sin^2(\theta_m/2) - \sin^2(\phi/2)}},\qquad 0 < \theta < \theta_m .\qquad(1.12)$$

From $t(0) = T/4$ it follows that the period T is

$$T = 2\sqrt{\frac{\ell}{g}}\int_0^{\theta_m}\frac{d\phi}{\sqrt{\sin^2(\theta_m/2) - \sin^2(\phi/2)}} .\qquad(1.13)$$

The integral in Eq. (1.13) can be calculated numerically. Alternatively, it can exactly be expressed in terms of K, the so-called complete elliptic integral of the first kind, see Appendix A. The answer is

$$T = 4\sqrt{\frac{\ell}{g}}K\left(\sin\frac{\theta_m}{2}\right).$$ (1.14)

Equation (1.14) can be derived from (1.13) by changing the variable $t = \sin(\phi/2)/\sin(\theta_m/2)$ and comparing to Eq. (A.16) of Appendix A, which is the definition of K; alternatively, we can look up the integral in Eq. (1.13) in standard integral tables such as [6], or use symbolic integration routines.

Equation (1.14) is exact for our simple pendulum model but, because of the special function, not directly informative. An expression for T that is quite revealing can be found for the important special case where θ_m is small: since $0 < \phi < \theta_m$ in Eq. (1.13), ϕ is also small. We can therefore approximate both sines in Eq. (1.13) by the first terms in their Maclaurin expansions, amounting to the replacement of $\sin^2(\theta_m/2) - \sin^2(\phi/2)$ by $(\theta_m^2 - \phi^2)/4$. This leads to the approximate expression

$$T \sim 4\sqrt{\frac{\ell}{g}}\int_0^{\theta_m}\frac{d\phi}{\sqrt{\theta_m^2 - \phi^2}},$$ (1.15)

in which the integral can be evaluated (exactly) by elementary means: its value is independent of θ_m and equals $\pi/2$. We are thus led to the important approximate expression

$$T \sim 2\pi\sqrt{\frac{\ell}{g}}.$$ (1.16)

In Eq. (1.16), the independence from θ_m means that small pendulum oscillations are "isochronous," a term originating from a Greek word meaning "equal in time."

An interesting interpretation of Eq. (1.16) can be obtained via a different derivation that proceeds from the equation of motion: if θ_m is small, then $\theta(t)$ is also small for all t, and we can replace $\sin\theta(t)$ by $\theta(t)$ in Eq. (1.8). This leads to the differential equation

$$\theta''(t) + \omega^2\theta(t) = 0, \qquad \text{where} \qquad \omega = \sqrt{\frac{g}{\ell}},$$ (1.17)

which is approximate for our model, but linear and elementary: its solution exactly satisfying the two initial conditions is

$$\theta(t) = \theta_m\cos(\omega t), \qquad \left(\omega = \sqrt{g/\ell}\right).$$ (1.18)

Thus, when the initial displacement is small, the oscillatory motion is "simple harmonic" with period $T = 2\pi/\omega$ and we are once again led to Eq. (1.16). The approximation $\sin\theta(t) \cong$

$\theta(t)$ amounts to approximating the restoring force by a quantity proportional to the arclength $s(t)$; a force of this type is generally associated with harmonic oscillations [5].

Because $K(0) = \pi/2$ (see Appendix A), our first way of showing Eq. (1.16) amounts to setting $\theta_m = 0$ in Eq. (1.14). This suggests a better than Eq. (1.16) approximation for small θ_m: the full Maclaurin expansion of $K(x)$ is available in handbooks and in Eq. (A.17) of Appendix A. We limit ourselves to the first two terms, i.e., to the small-x approximation $K(x) \sim (\pi/2)\,(1 + x^2/4)$. Eq. (1.14) thus gives

$$T \sim 2\pi\sqrt{\frac{\ell}{g}}\left(1 + \frac{1}{4}\sin^2\frac{\theta_m}{2}\right). \tag{1.19}$$

Although it is not necessary to do so, we can further replace $\sin^2(\theta_m/2)$ by $(\theta_m/2)^2$ to obtain

$$T \sim 2\pi\sqrt{\frac{\ell}{g}}\left(1 + \frac{\theta_m^2}{16}\right). \tag{1.20}$$

Equation (1.20) is an improvement to Eq. (1.16): the expression in Eq. (1.20) is a quadratic (rather than a linear and constant) function of θ_m, revealing that the period slightly increases if the small initial amplitude θ_m increases. Approximations (1.16) and (1.20) are shown in Fig. 1.5. Also shown is the "exact" value, found from Eq. (1.14) using a standard computer routine for the calculation of the elliptic integral. (Numerical integration of Eq. (1.13) gives identical results.) Corresponding percent (relative) errors are shown in Fig. 1.6 as the dashed and dot-dashed lines. Naturally, Eq. (1.20) is better than Eq. (1.16). The errors decrease with decreasing θ_m, just as expected. Furthermore, the errors are quite small, rising to 6.8% and 0.43% at the rightmost angle $\theta_m = 60°$.

The second term in Eq. (1.20), i.e., the quantity

$$r = 2\pi\sqrt{\frac{\ell}{g}}\frac{\theta_m^2}{16} \tag{1.21}$$

is, approximately, the absolute error (or remainder) of the cruder approximation in Eq. (1.16). The corresponding percent error $r/T \times 100$ (with T found from Eq. (1.14)) is also shown in Fig. 1.6 as the solid line; it is seen to be very close to the exact percent error (dashed line), especially for small θ_m.

1.3 A DIFFERENTIAL EQUATION

Our next example has to do with a differential equation. For $x > 0$, consider the linear, homogeneous, second-order equation

$$xf''(x) + (1 - x)f'(x) - f(x) = 0, \tag{1.22}$$

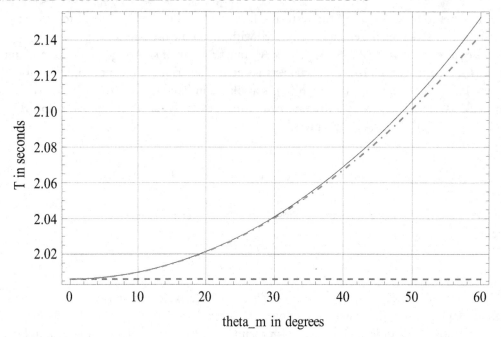

Figure 1.5: Exact (solid line), approximate (dashed line), and improved approximate (dot-dashed line) expressions for period T of a simple pendulum as calculated by (1.14), (1.16), and (1.20), respectively; $\ell = 1$ meter.

to be solved together with the condition

$$f(+\infty) = 0 . \tag{1.23}$$

By inspection, one solution to Eq. (1.22) is $f(x) = e^x$. We seek another solution satisfying Eq. (1.23).

1.3.1 "SERIES SOLUTION"

We first apply the well-known technique of seeking a series solution see, e.g., [7]. In view of Eq. (1.23), we attempt to find a solution in the form of a series of inverse powers of x,

$$f(x) = \sum_{n=1}^{\infty} \frac{\alpha_n}{x^n} . \tag{1.24}$$

In Eq. (1.24), where there is no constant term, the coefficients α_n are to be determined. We are thus seeking a solution satisfying Eq. (1.23) that is analytic at $x = \infty$.

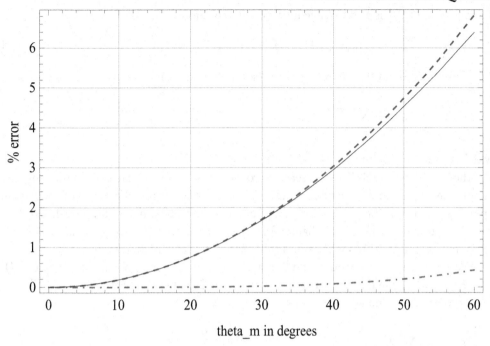

Figure 1.6: Relative (percentage) error between exact T and (i) approximate T of Fig. 1.5 (dashed line); and (ii) improved approximate T of Fig. 1.5 (dot-dashed line). Note that these relative errors are actually independent of ℓ. The solid line is $r/T \times 100$, with r defined in (1.21) and T calculated from (1.14).

Substitution into Eq. (1.22) and term-by-term differentiation yields

$$x \sum_{n=1}^{\infty} n(n+1)\alpha_n x^{-n-2} - (1-x) \sum_{n=1}^{\infty} n\alpha_n x^{-n-1} - \sum_{n=1}^{\infty} \alpha_n x^{-n} = 0 . \tag{1.25}$$

Shift indices to obtain

$$\sum_{n=2}^{\infty} (n-1)n\alpha_{n-1}x^{-n} - \sum_{n=2}^{\infty} (n-1)\alpha_{n-1}x^{-n} + \sum_{n=1}^{\infty} n\alpha_n x^{-n} - \sum_{n=1}^{\infty} \alpha_n x^{-n} = 0 . \tag{1.26}$$

Since the Maclaurin series on the left-hand side is zero, the coefficients of all powers x^{-n} of x^{-1} must be zero. For $n = 1$, we get an identity ($\alpha_1 - \alpha_1 = 0$); for the remaining values of n, we obtain $(n-1)^2\alpha_{n-1} + (n-1)\alpha_n = 0$ or

$$\alpha_n = -(n-1)\alpha_{n-1}, \qquad n = 2, 3, \ldots . \tag{1.27}$$

Equation (1.27) is a recurrence relation with solution

$$\alpha_n = (-1)^{n-1}(n-1)!\alpha_1, \quad n = 2, 3, \ldots ,\tag{1.28}$$

and we have thus found the following formal solution to Eq. (1.22),

$$f(x) = \sum_{n=1}^{\infty} \frac{(-1)^{n-1}(n-1)!}{x^n} = \frac{0!}{x} - \frac{1!}{x^2} + \frac{2!}{x^3} - \frac{3!}{x^4} + \cdots ,\tag{1.29}$$

in which we set $\alpha_1 = 1$ for brevity.

The "series" in Eq. (1.29) has a serious problem: because of the factorial in the numerator, it does not converge for any value of x, however large! In other words, the radius of convergence of the power series is zero (as can be proved by applying the well-known ratio test; this and related topics will be further discussed in Chapter 3). Checking the convergence of obtained power series is, of course, an essential step when one applies series-solution methods [7].

We were thus unsuccessful in finding a solution of the type postulated in Eq. (1.24). We can nevertheless ask whether the formal series in Eq. (1.29) can in some way be useful. The answer turns out to be yes, in the following sense: define $f_N(x)$ by

$$\begin{aligned}
f_N(x) = \sum_{n=1}^{N} \frac{(-1)^{n-1}(n-1)!}{x^n} &= \frac{0!}{x} - \frac{1!}{x^2} + \frac{2!}{x^3} - \frac{3!}{x^4} + \cdots \\
&+ \frac{(-1)^{N-1}(N-1)!}{x^N}, \quad N = 1, 2, \ldots .
\end{aligned}\tag{1.30}$$

As opposed to the $f(x)$ of Eq. (1.29), $f_N(x)$—which we can call a "partial sum" because of the finite summation limit—is a well-defined quantity for any (finite) value of N. While it is true that

$$\lim_{N\to\infty} f_N(x) \quad \text{does not exist for any} \quad x > 0 ,\tag{1.31}$$

in Section 1.3.2 below we will numerically verify that, for large x, *each* $f_N(x)$ well-approximates the solution to Eq. (1.22) and Eq. (1.23), with the approximation improving with increasing x.

1.3.2   THE SOLUTION AS AN INTEGRAL: INTEGRATION BY PARTS

To achieve this, we first consider Eqs. (1.22) and (1.23) from a different viewpoint. Having already observed that $f(x) = e^x$ satisfies Eq. (1.22), we set

$$f(x) = e^x g(x)\tag{1.32}$$

in Eq. (1.22) in order to obtain a solvable, first-order equation for $g'(x)$; this is the well-known method of "reduction of order" discussed, for example, in [7]. Differentiating Eq. (1.32), we get

$$f'(x) = e^x \left[g'(x) + g(x)\right] \quad \text{and} \quad f''(x) = e^x \left[g''(x) + 2g'(x) + g(x)\right] .\tag{1.33}$$

Substitution of Eqs. (1.32) and (1.33) into Eq. (1.22) yields

$$xg''(x) + (x+1)g'(x) = 0 .$$ (1.34)

As expected from the reduction-of-order method, Eq. (1.34) is a first order equation for $g'(x)$. It is easily solved by recasting as

$$[xg'(x)]' + xg'(x) = 0 ,$$ (1.35)

which gives $xg'(x) = C_1 e^{-x}$. Therefore, $g'(x) = C_1 e^{-x}/x$ and

$$g(x) = -C_1 \int_x^\infty \frac{e^{-t}}{t} dt + C_2 .$$ (1.36)

Combining Eq. (1.32) with Eq. (1.36), choosing $C_2 = 0$ so as to satisfy Eq. (1.23), and setting $C_1 = -1$ for brevity, we obtain

$$f(x) = e^x \int_x^\infty \frac{e^{-t}}{t} dt .$$ (1.37)

The function given by Eq. (1.37) (or any constant times that function) is an integral representation of the solution to Eqs. (1.22) and (1.23).

The integral in Eq. (1.37) is a special function, the so-called *exponential integral* $E_1(x)$, but we choose to ignore this for the time being. Instead, let us twice integrate by parts:

$$f(x) = -e^x \int_x^\infty \frac{1}{t} \frac{de^{-t}}{dt} dt = \frac{1}{x} - e^x \int_x^\infty \frac{e^{-t}}{t^2} dt = \frac{1}{x} - \frac{1}{x^2} + 2e^x \int_x^\infty \frac{e^{-t}}{t^3} dt .$$ (1.38)

Integrating by parts N times yields

$$f(x) = \frac{0!}{x} - \frac{1}{x^2} + \frac{2!}{x^3} - \ldots + \frac{(-1)^{N-1}(N-1)!}{x^N}$$
$$+ (-1)^N N! e^x \int_x^\infty \frac{e^{-t}}{t^{N+1}} dt, \quad N = 1, 2, \ldots .$$ (1.39)

In Eq. (1.39), we can recognize what we previously called the "partial sum" $f_N(x)$. We have thus found

$$f(x) = f_N(x) + (-1)^N N! e^x \int_x^\infty \frac{e^{-t}}{t^{N+1}} dt = f_N(x) + r_N(x), \quad N = 1, 2, 3, \ldots ,$$ (1.40)

where $f_N(x)$ is defined in Eq. (1.30). Equation (1.40) is useful because it is an explicit formula for the remainder $r_N(x)$—no such formula was obtained in Section 1.3.1.

Figure 1.7 shows the relative errors $r_1(x)/f(x)$, $r_2(x)/f(x)$, and $r_3(x)/f(x)$ as function of x, where $r_N(x)$ and $f(x)$ were found by numerically calculating the corresponding integrals in Eqs. (1.40) and (1.37). (Identical curves are obtained if one calculates $r_N(x)$ from $r_N(x) = f(x) - f_N(x)$ using Eq. (1.37) and the sum in Eq. (1.30).) For each value $N = 1, 2$, and 3, the errors in Fig. 1.7 are seen to decrease with increasing x, in accordance with our (as yet imprecise) definition of an asymptotic approximation. At the rightmost point $x = 15$ of Fig. 1.7, the errors for $N = 1, 2$, and 3 are -6.3%, 0.8%, and -0.15%, respectively. If x is increased to $x = 30$, the respective errors become even smaller, -3.2%, 0.2%, and -0.02%.

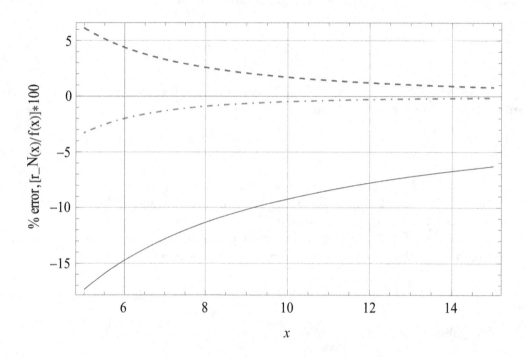

Figure 1.7: Relative (percentage) error $r_N(x)/f(x) \times 100$ between (i) exact $f(x)$ of (1.37) and (ii) approximation $f_N(x)$ of (1.30), as a function of x for $N = 1$ (solid line), $N = 2$ (dashed line), and $N = 3$ (dot-dashed line).

From Fig. 1.7, it is also seen that $f_1(x)$ and $f_3(x)$ overestimate $f(x)$, while $f_2(x)$ underestimates $f(x)$, a behavior readily explained through the factor $(-1)^N$ in Eq. (1.40). This behavior is unlike that of Fig. 1.5, in which both approximations underestimate the exact value.

For any fixed value of x in Fig. 1.7, it is seen that the best approximation is provided by $f_3(x)$, followed by $f_2(x)$, and finally by $f_1(x)$. This observation should not, however, lead us to think that we can improve the approximation as much as desired by indefinitely increasing N.

Indeed, this would imply the three equivalent statements $\lim_{N\to\infty} r_N(x)/f(x) = 0$, $\lim_{N\to\infty} r_N(x) = 0$, and $\lim_{N\to\infty} f_N(x) = f(x)$, contradicting Eq. (1.31).

The behavior of $f_N(x)$ for large N can be better understood from Fig. 1.8, which shows $|r_N(x)|/f(x)$ as function of N, for $x=5$ (dots) and $x=6$ (continuous line). As already apparent from Fig. 1.7, the error is always smaller (in absolute value) for the larger of the two values, namely for $x = 6$. What Fig. 1.8 further reveals is the divergence for large N: after an initial decrease, both errors eventually increase (and the increase continues for yet larger values of N). Accordingly, there is an optimal value of N for $x = 5$ and a different one for $x = 6$.

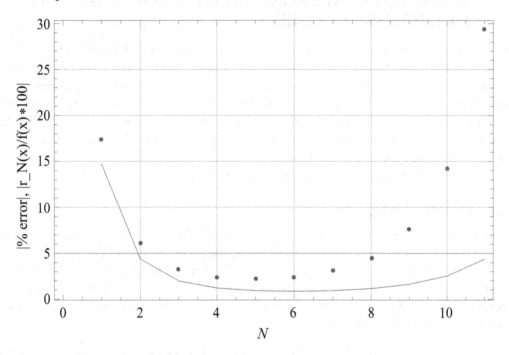

Figure 1.8: Absolute value of relative (percentage) error $|r_N(x)|/f(x) \times 100$ between (i) exact $f(x)$ of (1.37) and (ii) approximation $f_N(x)$ of (1.30), as a function of $N (N = 1, 2, ..., 11)$ for $x = 5$ (dots) and $x = 6$ (continuous line).

In the next chapter, we will revisit $f(x)$ after carefully defining the concept of an asymptotic approximation. We will show that $f_N(x)$ (as defined in Eq. (1.30)) is an asymptotic approximation of $f(x)$ (as defined in Eq. (1.37)) as $x \to \infty$, with the remainder small in an appropriate sense. We will denote this by

$$f(x) = \sum_{n=1}^{N} \frac{(-1)^{n-1}(n-1)!}{x^n} + O\left(\frac{1}{x^{N+1}}\right) \quad (x \to \infty). \qquad (1.41)$$

In Eq. (1.41), N is finite but arbitrary; this fact will be reflected by the notation

$$f(x) \sim \sum_{n=1}^{\infty} \frac{(-1)^{n-1}(n-1)!}{x^n} \quad (x \to \infty), \tag{1.42}$$

that we will also carefully define in the next chapter. We will say that the formal series on the right of Eq. (1.42) is an asymptotic expansion of the function on the left.

1.4   ASYMPTOTIC APPROXIMATIONS FOR HIGH-SWR TRANSMISSION LINES

Let $G(z) + jB(z)$ be the admittance along a lossless transmission line, normalized to the characteristic conductance of the line [8]. The dimensionless quantities $G(z)$ and $B(z)$ are, respectively, the normalized conductance and susceptance. For brevity, we omit the adjective "normalized" in what follows. Our purpose is to find asymptotic formulas [8] for $G(z)$ and $B(z)$ subject to the condition that the line is highly mismatched. These formulas supplement the Smith chart, help one understand the spatial variations of conductance and susceptance along the line, and are appropriate for quick, back-of-the-envelope calculations, especially in regions (to be called large-g regions below) where one cannot easily use the Smith chart.

1.4.1   EXACT FORMULAS

We assume a time dependence $e^{j\omega t}$, where $\omega = 2\pi f = 2\pi v/\lambda = v\beta$ and λ is the line wavelength, β is the phase constant, and v is the velocity of propagation. The line is driven by a generator at one end, loaded at the other end, and is assumed to be a number of wavelengths long. The coordinate z measures the distance from the load, with movement away from the load (and towards the generator) corresponding to increasing z. Let $G_{\max}$ or $G_{\min}$ denote the maximum and minimum value of conductance along the line,

$$G_{\max} = \max_z G(z), \quad G_{\min} = \min_z G(z). \tag{1.43}$$

It is well-known—and apparent from the Smith chart—that these two parameters are reciprocals of one another and related to the magnitude of the reflection coefficient $|\rho|$ and the standing-wave ratio SWR by

$$\frac{1 + |\rho|}{1 - |\rho|} = G_{\max} = \frac{1}{G_{\min}} = \text{SWR} \tag{1.44}$$

[8]. One can thus express $|\rho|$ in terms of any one of the quantities $G_{\max}$, $G_{\min}$, or SWR, e.g.,

$$|\rho| = \frac{G_{\max} - 1}{G_{\max} + 1}. \tag{1.45}$$

It is shown in [8] that, as we move along the line, the susceptance becomes zero when $z = z_0$, where

$$z_0 = \frac{\lambda}{4\pi}\text{Arctan}\left[\frac{2B(0)}{G^2(0) + B^2(0) - 1}\right] , \tag{1.46}$$

in which the inverse tangent function is multivalued. It is also shown in [8] that all zero susceptance points coincide with maximum or minimum conductance points. Thus, when $z = z_0$, we have $B(z_0) = 0$ together with either $G(z_0) = G_{\max}$ or $G(z_0) = G_{\min}$. This can be verified from the Smith chart.

Our analysis is greatly facilitated by choosing z_0 to be any maximum conductance point and shifting the origin to this point. In other words, we define $x = z - z_0$ and

$$g(x) = G(x + z_0) \quad \text{and} \quad b(x) = B(x + z_0) . \tag{1.47}$$

Under these circumstances, it is shown in [8] that the variations of conductance and susceptance along the line are governed by

$$g(x) = \frac{G_{\max}\left(1 + \tan^2\beta x\right)}{1 + G^2_{\max}\tan^2\beta x} , \tag{1.48}$$

$$b(x) = -\frac{(G^2_{\max} - 1)\tan\beta x}{1 + G^2_{\max}\tan^2\beta x} . \tag{1.49}$$

Equations (1.48) and (1.49) are exact (within the context of ordinary, lossless transmission-line theory). They are more concise than the more usual formulas for $G(z)$ and $B(z)$ (i.e., those expressed in terms of the distance from the load rather than from a maximum conductance point) [8], and are the cornerstones for the discussions that follow.

1.4.2   FURTHER EXACT FORMULAS: LARGE- AND SMALL- g REGIONS

We first discuss some exact consequences of Eqs. (1.48) and (1.49). From Eq. (1.48) we see that $g(x)$ attains the value 1 when $x = x_3$, where $\tan^2\beta x_3 = G_{\min} = 1/G_{\max}$. With Eq. (1.45), this last equation can be shown [8] to yield

$$x_3 = \frac{\lambda}{4\pi}\text{Arccos}\,|\rho| . \tag{1.50}$$

When $x = x_3$, it is seen from Eq. (1.49) that the susceptance attains one of the two values

$$b(x_3) = \pm\,(G_{\max} - 1)\,\sqrt{G_{\min}} . \tag{1.51}$$

There are regions in which $g(x)$ is large, and regions in which $g(x)$ is small. It is convenient to use the value 1 as a threshold and define the following two regions.

(i) Large-g regions: these consist of those points x on the line for which $1 < g(x) < G_{\max}$.

(ii) Small-g regions: here, $G_{\min} < g(x) < 1$.

In what follows, we suppose that $-\lambda/4 < x < \lambda/4$ (any other interval of width $\lambda/2$, centered at a maximum-conductance point can also be considered). There is one large-g region. It contains the maximum-conductance point $x = 0$, and its width is

$$\Delta x = 2x_3 = \frac{\lambda}{2\pi} \arccos |\rho| \ . \tag{1.52}$$

In Eq. (1.52) we use the principal value, arccos, of the multivalued function Arccos in accordance with the notation of the NIST *Digital Library of Mathematical Functions* [9]. The reader is invited to identify the large- and small- g regions on the Smith chart, and to compare values obtained from the exact formulas (1.50)–(1.52) to values obtained directly off the Smith chart.

We call a line "highly mismatched" if any one of the following equivalent conditions are satisfied

$$G_{\max} \gg 1, \quad \text{SWR} \gg 1, \quad G_{\min} \ll 1 \ . \tag{1.53}$$

For a highly mismatched line, we will examine the asymptotic behavior of $g(x)$ and $b(x)$ separately for the large- and small- g regions. Before doing this, we interpret the conditions in Eq. (1.53) in terms of the reflection coefficient: the reader is invited to use a Maclaurin expansion to show that, subject to Eq. (1.53), Eq. (1.45) implies

$$|\rho| \sim 1 - \frac{2}{G_{\max}} \ , \tag{1.54}$$

as well as the more precise asymptotic formula

$$|\rho| \sim 1 - \frac{2}{G_{\max}} + \frac{2}{G_{\max}^2} \ . \tag{1.55}$$

Equation (1.54) shows that the reflection coefficient is nearly 1, with a remainder equal, asymptotically, to $-2/G_{\max}$. We also say that this remainder is "of the order of" $1/G_{\max}$. Eq. (1.55) shows that the remainder of the asymptotic relation Eq. (1.54) is asymptotically equal to $2/G_{\max}^2$ and of the order of $1/G_{\max}^2$. While the meaning of our terminology should already be clear, we postpone precise definitions until the next chapter.

1.4.3   ASYMPTOTIC FORMULAS FOR THE LARGE-g REGION

(i) Let us first find an asymptotic formula for the width Eq. (1.52) of the large-g region: substitute Eq. (1.54) into Eq. (1.52) and approximate the resulting arccos by the first term in the identity

$$\arccos(1 - \varepsilon) = \sqrt{2\varepsilon} \left[1 + \sum_{k=1}^{\infty} \frac{1 \cdot 3 \cdot 5 \cdot \ldots \cdot (2k-1)}{2^{2k} (2k+1) \, k!} \varepsilon^k \right] , \qquad |\varepsilon| < 2 \ , \tag{1.56}$$

to obtain

$$\Delta x \sim \frac{\lambda}{\pi\sqrt{G_{\max}}}.$$ (1.57)

This shows that the region's width is of the order of $1/\sqrt{G_{\max}}$. Eq. (1.56)—which can be found in [9]—is, of course, the Maclaurin series of the function in the left-hand side, but the variable is $\sqrt{\varepsilon}$ rather than ε (the function is not analytic at $\varepsilon = 0$, and thus has no Maclaurin expansion in the variable ε). Alternatively, one can interpret the quantity within brackets in Eq. (1.56) as being the Maclaurin series of the analytic function $\arccos(1 - \varepsilon)/\sqrt{2\varepsilon}$. Such series will be further discussed in Chapter 3. For a proof of Eq. (1.56), see Problem 1.5.

(ii) In the large-g region, the following asymptotic formulas hold,

$$g(x) \sim \frac{G_{\max}}{1 + G_{\max}^2(\beta x)^2}, \qquad -\frac{\lambda}{2\pi\sqrt{G_{\max}}} < x < \frac{\lambda}{2\pi\sqrt{G_{\max}}},$$ (1.58)

$$b(x) \sim -\frac{G_{\max}^2(\beta x)}{1 + G_{\max}^2(\beta x)^2}, \qquad -\frac{\lambda}{2\pi\sqrt{G_{\max}}} < x < \frac{\lambda}{2\pi\sqrt{G_{\max}}}.$$ (1.59)

These are simple to derive: as long as we are in the large-g region, $\tan^2\beta x$ is small. In the numerator of Eq. (1.48), it can be neglected compared to 1. In the denominator, it is permissible to replace $\tan^2\beta x$ by the first term $(\beta x)^2$ in its Maclaurin expansion, thus leading to Eq. (1.58). Similarly, and also neglecting 1 compared to $G_{\max}^2$ in Eq. (1.49), one obtains Eq. (1.59). Note that the second term in the denominators of Eq. (1.58) and Eq. (1.59) can be small (when $|x|$ is very small) or large (when $|x|$ is near $\lambda/(2\pi\sqrt{G_{\max}})$).

(iii) At the ends $= \pm\Delta x/2$ of the large-g region, one has $g(\pm\Delta x/2) = 1$ so that the conductance is no longer large. The susceptance, however, is still large (as the reader can verify from the Smith chart). Its values are asymptotically given by

$$|b(\pm\Delta x/2)| \sim \sqrt{G_{\max}}.$$ (1.60)

Eq. (1.50) is a simple consequence of Eqs. (1.57) and (1.59).

(iv) Within the large-g region, $|b(x)|$ twice attains a maximum. The two points at which this occurs are:

$$x_1 \sim -\frac{\lambda}{2\pi G_{\max}} \quad \text{and} \quad x_2 \sim \frac{\lambda}{2\pi G_{\max}}.$$ (1.61)

This is seen by setting the derivative of Eq. (1.59) equal to zero.

(v) At the points x_1 and x_2, the susceptance values are equal (or opposite) to the conductance values, which are half the maximum conductance values, viz.

$$b(x_1) = -b(x_2) \sim g(x_1) = g(x_2) \sim \frac{1}{2}G_{\max}.$$ (1.62)

The asymptotic relations in Eq. (1.62) are simple consequences of Eqs. (1.58), (1.59), and (1.61).

Equation (1.62) provides a second method for measuring the width of the narrow conductance peak, by taking the distance between the two points where the conductance drops to half its maximum (peak) value G_{max}. If this distance is δx, then, by Eq. (1.61),

$$\delta x = x_2 - x_1 \sim \frac{\lambda}{\pi G_{max}} \, . \tag{1.63}$$

The width δx of the peak is therefore of the order of $1/G_{max}$. For sufficiently large G_{max}, the distance δx is, of course, much smaller than the width Δx.

Formulas (1.57)–(1.63) hold within the large-g region of a highly mismatched line. Put differently, they hold near the rightmost point of the Smith chart. As we move closer to this point, it becomes increasingly difficult to read accurate values off the Smith chart, see Problem 1.6. By contrast, the accuracy of Eqs. (1.57)–(1.63) continues to increase.

It is likely that the reader has noticed analogies of the spatial behavior of the conductance and susceptance (discussed herein) to the frequency behavior of the input conductance/susceptance in a high-Q, series RLC circuit. For example, δx is analogous to the 3db-bandwidth. Further such analogies are discussed in [8].

1.4.4   ASYMPTOTIC FORMULAS FOR THE SMALL-g REGION

Simplified formulas can also be obtained for the small-g region, which is wider of the two regions: by Eq. (1.57), $\tan \beta x$ is at most of the order of $1/\sqrt{G_{max}}$. Therefore, one can neglect the 1 in the denominator of Eq. (1.48) to obtain:

$$g(x) \sim \frac{G_{min}}{\sin^2 \beta x}, \quad \frac{\lambda}{2\pi \sqrt{G_{max}}} < |x| < \frac{\lambda}{4} \, . \tag{1.64}$$

In Eq. (1.49), neglect also the 1 in the numerator. The result is

$$b(x) \sim - \cot \beta x \, , \quad \frac{\lambda}{2\pi \sqrt{G_{max}}} < |x| < \frac{\lambda}{4} \, . \tag{1.65}$$

Equation (1.65) reveals that, within the small-g region of a highly mismatched line, the susceptance values are roughly independent of the load. The reader is invited to verify this on the Smith chart.

1.5   SUPPLEMENTARY REMARKS AND FURTHER READING

The simple pendulum of Section 1.2 (also called "simple gravity pendulum" and "plane pendulum" because the motion is confined to be within a plane) is discussed in most elementary textbooks on mechanics, physics, and dynamics. An explicit solution to Eq. (1.8)—and to its equivalents Eqs. (1.9) and (1.10)—can be found in terms of a related to K special function, the Jacobian elliptic function sn [9], [10]; to do this, one proceeds from Eq. (1.12) [10]. The "seminal obser-vation" [11] of the isochronous property was made by Galileo sometime between 1580 and 1590. According to Galileo's earliest biographer, the observation was made in the cathedral at Pisa, with Galileo timing the oscillations of chandeliers with his pulse, but this story may be apoc-ryphal [11]. Eq. (1.16) was discovered by Christiaan Huygens in 1673 [5]; Huygens's important contributions to the development of the pendulum clock are described in [11]. The easily solv-able differential equation Eq. (1.17) is often said to describe the "linearized pendulum;" this can be considered as an analog of many diverse situations [11], including the zero-input response of electric circuits such as the LC-circuit. In that case, Eq. (1.17) is satisfied by the capacitor charge with $\omega = 1/\sqrt{LC}$ being the resonant frequency.

The procedure of Section 1.3.1 amounts to first setting $y = 1/x$ and $h(y) = f(1/y)$, and then expanding about $y = 0$. Eqs. (1.22) and (1.23) become

$$y^3 h''(y) + y(1 + y)h'(y) - h(y) = 0 \tag{1.66}$$

and

$$h(0) = 0 . \tag{1.67}$$

What we did in Section 1.3.1 is the same as substituting

$$h(y) = \sum_{n=1}^{\infty} \alpha_n y^n \tag{1.68}$$

into Eq. (1.66) to find α_n in terms of α_1, and then realizing that the series we found has a zero radius of convergence.

In the language of the well-known theory of Frobenius and Fuchs [12, 13, 14], the expan-sion point $y = 0$ is an irregular singular point of Eq. (1.66), which is equivalent to saying that $x = \infty$ is an irregular singular point of Eq. (1.22). Therefore, there is no *a priori* reason to end up with a (convergent) solution of the type postulated in Eq. (1.68) (or, more generally, of the type $h(y) = y^s \sum_{n=0}^{\infty} \beta_n y^n$). Comprehensive discussions along these lines can be found in [12].

As already mentioned, we will later (in Chapter 2 and again in Problem 4.5 of Chapter 4) demonstrate that the divergent series we found in Section 1.3 is an asymptotic series. But we will not deal further with finding divergent series that are asymptotic to solutions of differential equations. For this topic, the interested reader is referred to [13], or to the briefer treatment in [9].

1.6   PROBLEMS

1.1.   Set $\varepsilon = z'/r$ in Eq. (1.3) and use a Maclaurin expansion to show Eq. (1.4). Improve (1.4) by finding the next term. How is that term connected to the error of the approximation in (1.4)?

1.2. In Section 1.2, we derived (1.18) by approximating the left-hand side of Eq. (1.8). Alternatively, one can show Eq. (1.18) by approximating the right-hand side of Eq. (1.10). Carry this out.

1.3. Find the next two terms in Eqs. (1.19) and (1.20).

1.4. Show Eqs. (1.54) and (1.55).

1.5. The multivalued function Arccos is defined by [9]

$$\text{Arc}\cos z = \int_z^1 \frac{dt}{\sqrt{1-t^2}} . \tag{1.69}$$

Use this definition to derive Eq. (1.56). *Hint:* Use Eq. (3.27) of Chapter 3.

1.6. A resonant (i.e., zero driving-point susceptance) antenna is the load of a lossless transmission line. The antenna (load) conductance is such that SWR=30. Determine the input admittance at a distance $x = 10.498\lambda$ from a zero susceptance point (i) by using exact formulas from Section 1.4; and (ii) by using asymptotic formulas from Section 1.4. Compare your answers. Can a reliable answer be obtained from the Smith chart?

REFERENCES

[1] J. M. Borwein and R. E. Crandall, "Closed Forms: What They Are and Why We Care," *Notices of the AMS*, vol. 60, no. 1, pp. 50–65, January 2013. DOI: 10.1090/noti936. 1

[2] C. A. Balanis, *Antenna Theory, Analysis and Design, 3rd Ed.* New York: Wiley, 2005, §3.4, § 4.2.4, § 4.5.2, and Chapter 6. 1, 3, 4

[3] R. W. P. King, G. Fikioris, and R. B. Mack, *Cylindrical Antennas and Arrays.* Cambridge, UK, Cambridge University Press, 2002, §1.6, §2.9. DOI: 10.1017/CBO9780511541100. 3

[4] C. A. Balanis, *Advanced Engineering Electromagnetics.* New York: Wiley, 1989, §1.4 and Chapter 13. 4

[5] J. B. Marion and S. T. Thornton, *Classical Dynamics of Particles and Systems, 4th Ed.* Dumfries, NC: Holt, Rinehart & Winston, 1995, Section 4.4. 5, 7, 19

[6] A. Jeffrey and D. Zwillinger, Eds., *Gradshteyn and Ryzhik's Table of Integrals, Series, and Products, 7th Ed.* New York: Academic Press, 2007. 6

[7] W. E. Boyce and R. C. Diprima, *Elementary Differential Equations and Boundary Value Problems, 9th Ed.* New York: Wiley, 2009, Chapters 3 and 5. 8, 10

[8] G. Fikioris, "Analytical studies supplementing the Smith Chart," *IEEE Trans. Education*, vol. 47, no. 2, pp. 261–268, May 2004. DOI: 10.1109/TE.2004.825512. 14, 15, 18

[9] F. W. J. Olver, D. W. Lozier, R. F. Boisvert, and C. W. Clark, *Digital Library of Mathematical Functions*, National Institute of Standards and Technology from `http://dlmf.nist.gov/` , §2.7, §4.23, §4.23.2, §4.24.2, §22.19.2. 16, 17, 19, 20

[10] D. F. Lawden, *Elliptic Functions and Applications*. New York: Springer, 1989, §5.1. DOI: 10.1007/978-1-4757-3980-0. 19

[11] G. L. Baker and J. A. Blackburn, *The Pendulum; a Case Study in Physics*. Oxford, UK: Oxford University Press, 2005. 19

[12] C. M. Bender and S. A. Orszag, *Advanced Mathematical Methods for Scientists and Engineers; Asymptotic Methods and Perturbation Theory*. New York: Springer, 1999, Chapter 3. DOI: 10.1007/978-1-4757-3069-2. 19

[13] F. W. J. Olver, *Asymptotics and Special Functions*. Natick, MA: A. K. Peters, 1997, Chapter 7. 19

[14] A. D. Polyanin and V. F. Zaitsev, *Handbook of Exact Solutions for Ordinary Differential Equations, 2nd Ed.* Boca Raton, FL: Chapaman & Hall/CRC, 2002, §0.2.2. DOI: 10.1201/9781420035339. 19

CHAPTER 2

Asymptotic Approximations Defined

In Chapter 1, we presented a number of approximations for simple problems. By nature of their derivations, these approximations improve as some problem parameters became larger or smaller, something that loosely justifies our calling them asymptotic. In the present chapter, we give the rigorous definition of an asymptotic approximation. We also define related concepts, such as that of order (already encountered in Chapter 1). We finally define an asymptotic expansion (a more general concept than an asymptotic approximation, encountered in Section 1.3). We illustrate all definitions by further examples.

2.1 DEFINITIONS

Let $f(x)$ and $g(x)$ be real- or complex-valued functions of the real variable x, and let $x_0 \in \mathbb{R}, x_0 = +\infty$, or $x_0 = -\infty$.

Definition 2.1 Asymptotic approximation.

The function $g(x)$ is an asymptotic approximation to $f(x)$ as $x \to x_0$, if $\lim\limits_{x \to x_0} \frac{f(x)}{g(x)} = 1$. We then say that $f(x)$ and $g(x)$ are asymptotic to each other as $x \to x_0$ and denote this by $f(x) \sim g(x) \, (x \to x_0)$.

Definition 2.2 Order.

The function $f(x)$ is big-oh of $g(x)$ as $x \to x_0$, or $f(x)$ is of the order of $g(x)$ as $x \to x_0$, if $\left| \frac{f(x)}{g(x)} \right|$ is bounded as $x \to x_0$. We denote this by $f(x) = O(g(x)) \, (x \to x_0)$.

Definition 2.3 Little-Oh.

The function $f(x)$ is much smaller than $g(x)$ as $x \to x_0$ or $f(x)$ is little-oh of $g(x)$ as $x \to x_0$, if $\lim\limits_{x \to x_0} \frac{f(x)}{g(x)} = 0$. This is denoted by $f(x) = o(g(x)) \, (x \to x_0)$ or $f(x) \ll g(x) \, (x \to x_0)$.

2.2 REMARKS AND EXAMPLES

- For finite x_0, we often limit ourselves to values of x larger (or smaller) than x_0. For example, the meaning of $f(x) \sim g(x) \, (x \to x_0^+)$ is given in Definition 2.1, but with a one-sided limit.

- Our definitions also hold for an integer variable x, with $x_0 = +\infty$, or $x_0 = -\infty$. See Problem 2.1 for an elementary example.

- Our three definitions relate the behavior of $f(x)$ to that of $g(x)$ as $x \to x_0$. There is no restriction on the behavior of $f(x)$ itself: $\lim_{x \to x_0} f(x)$ may be zero, finite, infinite, or nonexistent.

- If $f(x)$ is much smaller than $g(x)$ as $x \to x_0$, it follows from Definition 2.2 that $f(x)$ is of the order of $g(x)$ as $x \to x_0$. The converse is not true.

- The meaning of the often-used relations $f(x) = O(1) \, (x \to x_0)$ and $f(x) = o(1) \, (x \to x_0)$ is obvious from our more general definitions. The first means that $|f(x)|$ is bounded, and the second that the limit of $f(x)$ is zero.

- The relation $f(x) \ll g(x) \, (x \to x_0)$ may hold even if $f(x) > 0$, $g(x) < 0$, e.g., $x^4 \ll -x^2 (x \to 0)$.

- From our definitions, it is easily seen that $f(x) \sim g(x) \, (x \to x_0)$ is equivalent to $f(x) - g(x) \ll g(x) \, (x \to x_0)$, i.e., to the statement that the remainder $f(x) - g(x)$ is much smaller than the approximation $g(x)$ as $x \to x_0$. It is also equivalent to saying that the relative error $|f(x) - g(x)|/|g(x)|$ (or the relative error $|f(x) - g(x)|/|f(x)|$) converges to zero as $x \to x_0$. This is what we checked, numerically, in Figs. 1.6 and 1.7. A further equivalent relation is $f(x) = g(x) + o(g(x)) \, (x \to x_0)$.

- The main purpose of this book is to provide certain basic tools enabling us to find a simple function $g(x)$ such that $f(x) \sim g(x) \, (x \to x_0)$ for a given x_0 and a given, complicated function $f(x)$. Besides the examples already presented in Chapter 1, let us give one involving a special function. For $K_0(x)$, the modified Bessel function of order zero, it is true that

$$K_0(x) \sim \sqrt{\frac{\pi}{2}} \frac{e^{-x}}{x^{1/2}} \quad (x \to +\infty) . \tag{2.1}$$

- Besides finding simple asymptotic approximations to complicated functions, it is often useful to additionally determine the order of the remainder. We thus seek simple functions $g_1(x)$, $g_2(x)$ satisfying $f(x) = g_1(x) + O(g_2(x))$ and $g_2(x) = o(g_1(x)) \, (x \to x_0)$. These two relations, of course, imply $f(x) \sim g_1(x) \, (x \to x_0)$. For $K_0(x)$, for example, it is true that

$$K_0(x) = \sqrt{\frac{\pi}{2}} \frac{e^{-x}}{x^{1/2}} + O\left(\frac{e^{-x}}{x^{3/2}}\right) \quad (x \to +\infty) , \tag{2.2}$$

with Eq. (2.1) following immediately from Eq. (2.2).

- Here are some even more elementary asymptotic approximations:

$$(x+1)^3 \sim x^3, \quad (x+1)^3 = x^3 + o(x^3), \quad (x+1)^3 = x^3 + O(x^2) \ (x \to +\infty) . \quad (2.3)$$

The first two relations tell us that x^3 is an asymptotic approximation to $(x+1)^3$. According to the discussions above, the second relation gives us no further information than the first. The third additionally gives us the order of the remainder.

- In the previous example, $x_0 = +\infty$. Elementary examples for the case $x_0 = 0$ are

$$\frac{1}{1-x} \sim 1+x, \quad \frac{1}{1-x} = 1+x+o(x), \quad \frac{1}{1-x} = 1+x+O(x^2) \quad (x \to 0) . \quad (2.4)$$

Here, the approximation $1+x$ consists of two terms. The last statement gives the order of the remainder and is the strongest of the three.

- For large, positive values of x, the function $\sinh x$ is approximated by $\frac{1}{2}e^x$. This approximation is asymptotic and extremely accurate because the relative error is exponentially small, which amounts to saying that the remainder is exponentially smaller than the approximation itself, $\sinh x = \frac{1}{2}e^x + O(e^{-x}) \ (x \to +\infty)$. Let us stress that Eq. (2.2) tells us that the remainder is algebraically (and not exponentially) smaller than the approximation.

- Definition 2.1 immediately suggests a numerical check as to whether $f(x)$ and $g(x)$ are asymptotic as $x \to x_0$: one simply checks if the ratio of the two functions approaches 1 as x approaches x_0, or if the two functions become closer as x approaches x_0. This is precisely what we did in Fig. 1.5 (there, $f(x)$ is the T of Eq. (1.14), $g(x)$ is the T of Eq. (1.16)—or the T of Eq. (1.20)—with the roles of x and x_0 played by θ_m and 0, respectively). Such a check, of course, presupposes that one can numerically calculate both $f(x)$ and $g(x)$, something not always possible in practical situations.

- Asymptotic approximations can be multiplied in the sense that $f_1(x) \sim g_1(x)$ and $f_2(x) \sim g_2(x) \ (x \to x_0)$ implies $f_1(x)f_2(x) \sim g_1(x)g_2(x) \ (x \to x_0)$. This is a direct consequence of Definition 2.1. As we will soon see, addition is not as straightforward.

- **Powers and exponential functions.** For any value of λ, the obvious relations $x^\lambda = o(e^x)$ and $e^{-x} = o(x^\lambda) \ (x \to \infty)$ amount to the well-known fact that the exponential function e^x (e^{-x}) converges to infinity (converges to zero) faster than any power of x. It is also obvious that there is no λ for which $e^x = O(x^\lambda) \ (x \to \infty)$. If $f(x) \sim x^\lambda \ (x \to \infty)$ for some λ, then it is also true that $f(x) + e^{-x} \sim x^\lambda \ (x \to \infty)$. This demonstrates that two different functions can have the same asymptotic approximation.

- **Addition of Asymptotic Approximations.** The example above also shows that we must be careful when we sum asymptotic approximations: the difference between the last two relations has no meaning because, by Definition 2.1, no function can be asymptotic to zero. Conditions permitting addition are given in Problems 2.2 and 2.3.

- **Differentiation and Integration of Asymptotic Approximations.** As a rule, asymptotic approximations can be integrated. Differentiation, however, usually requires additional conditions, as shown by the example $x + \cos x \sim x$ $(x \to \infty)$: it is not true that $1 - \sin x \sim 1(x \to \infty)$. For these matters, the interested reader is referred to [1] and [2].

- **Maclaurin series-Taylor series.** In the simple examples of Chapter 1, we extensively used Maclaurin series. It is thus not surprising that truncated Maclaurin series form asymptotic approximations. More precisely, let $f(x) = \sum\limits_{n=0}^{\infty} \alpha_n x^n$. Then the partial sum $\sum\limits_{n=0}^{N} \alpha_n x^n$ is, for sufficiently small x, an approximation of $f(x)$ and the remainder equals the tail $\sum\limits_{n=N+1}^{\infty} \alpha_n x^n$, which is a convergent series. Clearly, the approximation is asymptotic (unless $\alpha_0, \alpha_1, \ldots, \alpha_N$ are all zero; as already noted, no function can be asymptotically approximated by zero). Furthermore, the remainder is $O(x^{N+1})$, i.e., of the order of the first neglected term; this important property will be further discussed below. These remarks are also true for Taylor series about any point x_0, including $x_0 = \infty$.

- **Pendulum of Section 1.2 revisited.** In Section 1.2, the quantities in the right-hand sides of Eqs. (1.16) and (1.20) are asymptotic approximations, as $\theta_m \to 0$, to the exact period T of Eq. (1.14) because they are truncated Maclaurin series. In Fig. 1.6, the increasing closeness, as $\theta_m \to 0$, of the solid line to the dashed line indicates that the remainder of the approximation in Eq. (1.16) is of the order of the first neglected term, i.e., the last term in Eq. (1.20).

- For problems involving more than one parameter, care must be taken before choosing the proper limits to work with: our definitions make sense when a single parameter approaches ∞ (or 0, or x_0, as the case may be). As a trivial example, we have $x/y = O(x)$ as $x \to \infty$ with y fixed, but $x/y = O(1)$ as $x \to \infty$ and $y \to \infty$ with x/y fixed, and $x/y = O(1/x)$ as $x \to \infty$ and $y \to \infty$ with x^2/y fixed. As another example, many large-x asymptotic approximations of the Bessel function $J_\nu(x)$ can be found in the literature; included are the two cases $x \to \infty$ (ν fixed), and $x \to \infty$ and $\nu \to \infty$ (x/ν fixed). It is not obvious beforehand which of the two would give a better numerical approximation to, say, $J_5(10)$.

- **Integral of Section 1.3.2 revisited.** In Section 1.3.2 we encountered the function $f(x)$ defined by the integral Eq. (1.37). Let us now demonstrate that each $f_N(x)$ defined in Eq. (1.30) is an asymptotic approximation, as $x \to \infty$, of $f(x)$, and estimate the order of the remainder. By Eq. (1.31), we are not dealing with a truncated Maclaurin series. In other words, it is incorrect to think that the remainder $r_N(x)$ equals the tail because the series $\sum_{n=N+1}^{\infty} \alpha_n x^n$ is divergent. But we can show

$$f(x) = f_N(x) + O\left(\frac{1}{x^{N+1}}\right) \quad (x \to \infty) \tag{2.5}$$

because we have available the explicit formula Eq. (1.40) for $r_N(x)$ (in practical situations, one does not always have such an explicit formula). From Eq. (1.40),

$$|r_N(x)| = N! e^x \int_x^{\infty} \frac{e^{-t}}{t^{N+1}}\, dt < \frac{N! e^x}{x^{N+1}} \int_x^{\infty} e^{-t}\, dt = \frac{N!}{x^{N+1}}, \quad N = 1, 2, \ldots . \tag{2.6}$$

Thus,

$$\frac{|r_N(x)|}{1/x^{N+1}} \underset{x \to \infty}{=} \text{bounded}, \quad N = 1, 2, \ldots . \tag{2.7}$$

For any finite value of N, we have thus analytically shown Eq. (1.41) and, in particular, that the remainder is of the order of the first neglected term.

2.3   COMPOUND ASYMPTOTIC APPROXIMATIONS

In this section, we discuss an inadequacy of our previous definitions.

2.3.1   ELEMENTARY EXAMPLE

It is true that $\sqrt{x^2 + 1} \sim x \, (x \to +\infty)$. Thus, the graphs of $\sin \sqrt{x^2 + 1}$ and $\sin x$ approach each other as x becomes larger, as we show in Fig. 2.1. Nevertheless, $\sin x$ is *not* an asymptotic approximation to $\sin \sqrt{x^2 + 1}$ as $x \to +\infty$: if it were, Definition 2.1 would require the two functions to have, for sufficiently large x, *exactly* the same zeros. While the large zeros increasingly approach one another, they do not coincide. This can already be observed for the values of x in Fig. 2.1. The proximity of the large zeros can in fact be quantified using asymptotics (Problem 2.1). In any case, our definition—as it stands—seems incapable of capturing the closeness of the two graphs exhibited in Fig. 2.1.

While one should know about this difficulty, we agree with Bender and Orszag that one should not consider it major: In their well-known book [3], Bender and Orszag give many useful examples of asymptotic relations before discussing the difficulty. They then remark, "After all the successful asymptotic analysis of the previous three sections, it is surprising to encounter such a silly flaw in the definition of an asymptotic relation," and observe that relations such as

$$\sin \sqrt{x^2 + 1} \sim \sin x \, (x \to +\infty), \tag{2.8}$$

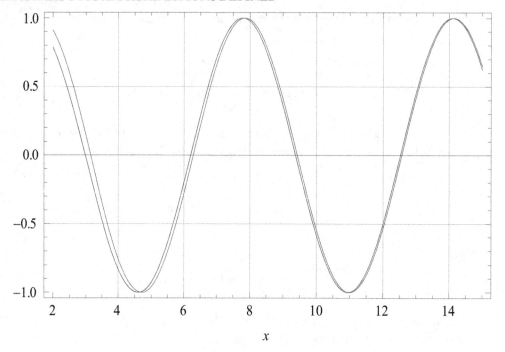

Figure 2.1: $\sin x$ and $\sin \sqrt{x^2 + 1}$ for $2 < x < 15$.

"want to be valid asymptotic relations." For examples such as the one discussed here (but not in all cases), one can avoid writing relations like Eq. (2.8) by using "compound asymptotic approximations," a concept we now proceed to illustrate.

2.3.2   BESSEL FUNCTION OF ORDER ZERO

The well-known Bessel function of order zero possesses the following integral representation [1],

$$J_0(x) = \frac{2}{\pi} \int_1^\infty \frac{\sin yx}{\sqrt{y^2 - 1}} dy, \quad x > 0 .$$ (2.9)

The integral in Eq. (2.9) is a Fourier sine transform. Change the variable $t = y - 1$ to obtain

$$J_0(x) = \frac{2}{\pi} f_1(x) \sin x + \frac{2}{\pi} f_2(x) \cos x, \quad x > 0 ,$$ (2.10)

where

$$f_1(x) = \int_0^\infty \frac{\cos tx}{\sqrt{t}\sqrt{t + 2}} dt \quad \text{and} \quad f_2(x) = \int_0^\infty \frac{\sin tx}{\sqrt{t}\sqrt{t + 2}} dt .$$ (2.11)

In Chapter 4 we will show that, in the sense of Definition 2.1,

$$f_1(x) \sim \sqrt{\frac{\pi}{4x}} \quad \text{and} \quad f_2(x) \sim \sqrt{\frac{\pi}{4x}} \quad (x \to \infty) . \tag{2.12}$$

Equations (2.10) and (2.12) suggest that for large values of x, $J_0(x)$ well approximates the function $\frac{2}{\pi}\sqrt{\frac{\pi}{4x}}(\sin x + \cos x) = \sqrt{\frac{2}{\pi x}}\cos(x - \frac{\pi}{4})$. This is verified in Fig. 2.2; here, the two graphs happen to be close even for relatively small values of x.

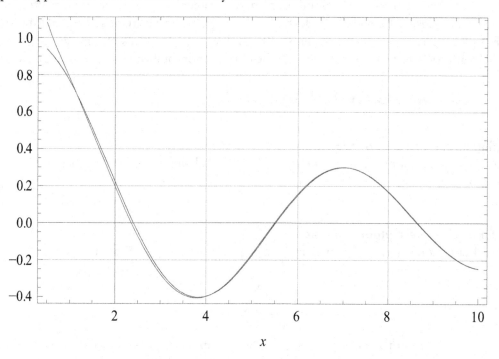

Figure 2.2: $J_0(x)$ and $\sqrt{\frac{2}{\pi x}}\cos\left(x - \frac{\pi}{4}\right)$ for $1/2 < x < 10$.

Despite the closeness for large values of x, the two functions $J_0(x)$ and $\sqrt{\frac{2}{\pi x}}\cos(x - \frac{\pi}{4})$ do not exactly have the same zeros. Therefore, as we explained in Section 2.3.1, it is strictly incorrect to write

$$J_0(x) \sim \sqrt{\frac{2}{\pi x}}\cos\left(x - \frac{\pi}{4}\right) \quad (x \to \infty) . \tag{2.13}$$

One should contrast Eq. (2.13) with the (exact) *equality* Eq. (2.10) and the *two asymptotic approximations* Eq.(2.12); these three relations form a "compound asymptotic approximation." In Problem 2.6, the reader is asked to find a compound asymptotic approximation that can replace relation Eq. (2.8) of our previous (more elementary) example.

We have seen that the valid relations $(2/\pi)f_1(x)\sin x \sim (2/\pi)\sqrt{\pi/(4x)}\sin x$ and $(2/\pi)f_1(x)\cos x \sim (2/\pi)\sqrt{\pi/(4x)}\cos x$ cannot be added to obtain a valid (in the sense of Definition 2.1) asymptotic approximation. We can view this as a counterexample that shows that it is not always possible to add asymptotic approximations. But we should not stop here: Eq. (2.13)—which can be shown using independent methods—is very useful, and often referred to as the large-argument asymptotic approximation of $J_0(x)$. In many problems, it is often necessary to approximate complicated oscillatory functions using sines and cosines. One sometimes uses nonrigorous methods that result in relations like Eq. (2.13) and it may not be a simple matter to find remedies like Eqs. (2.10) and (2.12). In such cases, one can sacrifice rigor and regard expressions like Eq. (2.13) as true asymptotic approximations. Frequently, the lack of rigor can be compensated by after-the-fact checks, e.g., through the physics of the problem, or numerical calculations.

2.4   ASYMPTOTIC EXPANSIONS

Definition 2.4   Asymptotic Sequence.
A sequence of functions $\psi_0(x), \psi_1(x), \ldots$ is called an asymptotic sequence as $x \to x_0$ if, for all n, $\psi_{n+1}(x) \ll \psi_n(x)$ $(x \to x_0)$.

Definition 2.5   Asymptotic Expansion.
Let $\psi_0(x), \psi_1(x), \ldots$ be an asymptotic sequence as $x \to x_0$. We write

$$f(x) \sim \sum_{n=0}^{\infty} \alpha_n \psi_n(x) \tag{2.14}$$

to denote that the formal series $\sum_{n=0}^{\infty} \alpha_n \psi_n(x)$ is an asymptotic expansion of $f(x)$ as $x \to x_0$. This means that

$$f(x) = \sum_{n=0}^{N} \alpha_n \psi_n(x) + O\left(\psi_{N+1}(x)\right) \tag{2.15}$$

for all $N = 0, 1, 2, \ldots$.

An asymptotic expansion like Eq. (2.14) is also called an asymptotic expansion of the Poincaré type [4]. Equation (2.15) means that, for all finite N, the remainder is of the order of the first neglected term. The term with $n=0$ in Eq. (2.15) is called the dominant term.

It is clear from our previous discussions that the Taylor series of a given function $f(x)$ about $x = x_0$ is an asymptotic expansion of $f(x)$ as $x \to x_0$, with $\psi_n(x) = (x - x_0)^n$ if x_0 is finite, or $\psi_n(x) = 1/x^n$ if $x_0 = \infty$. Taylor series are convergent series but, as we have already noted, an asymptotic series need not be convergent.

For $x_0 = 0^+$, an often-encountered example of an asymptotic sequence is $\psi_0(x) = \ln x$, $\psi_1(x) = 1$, $\psi_2(x) = x \ln x$, $\psi_3(x) = x$, $\psi_4(x) = x^2 \ln x$, $\psi_5(x) = x^2$,.... The small-argument asymptotic expansion of $K_0(x)$, for instance, is [1]

$$K_0(x) \sim -\ln x + (\ln 2 - \gamma) - \frac{1}{4} x^2 \ln x + \frac{\ln 2 + 1 - \gamma}{4} x^2 + \ldots \quad (x_0 \to 0^+) , \quad (2.16)$$

where $\gamma = 0.577216\ldots$ is Euler's constant, see Chapter 3. For clarity, only the first few terms appear in Eq. (2.16).

For $x_0 = +\infty$, $\psi_n(x) = 1/x^n$ is the most usual asymptotic sequence; the corresponding asymptotic expansion is called an asymptotic power series, or a Poincaré asymptotic expansion (or, more simply, an asymptotic expansion—see Section 2.5 for historical remarks). Such is the asymptotic expansion of the function $f(x)$ of Section 1.3: we have already shown that the remainder is of the order of the first neglected term, see Eq. (2.6). For $x_0 = +\infty$, another usual asymptotic sequence is $\psi_n(x) = e^{rx} x^s / x^n$, where r and s are independent of n. For example, the asymptotic expansion of $K_0(x)$ is of this type with $r = -1$ and $s = -1/2$ [1]:

$$K_0(x) \sim \sqrt{\frac{\pi}{2}} \frac{e^{-x}}{x^{1/2}} \left[1 + \sum_{n=1}^{\infty} (-1)^n \frac{1^2 \cdot 3^2 \cdot 5^2 \cdots (2n-1)^2}{n! 8^n x^n} \right] \quad (x \to +\infty) . \quad (2.17)$$

Both Eqs. (2.1) and (2.2) are immediate consequences of (2.17). Note that the symbol $\sim$ is used both for asymptotic approximations and asymptotic expansions.

Let $\psi_0(x), \psi_1(x), \ldots$ and $\xi_0(x), \xi_1(x), \ldots$ be two asymptotic sequences as $x \to x_0$. We say that $f(x)$ has a compound asymptotic expansion if

$$f(x) = \gamma(x) f_1(x) + \delta(x) f_2(x) \quad (2.18)$$

with

$$f_1(x) \sim \sum_{n=0}^{\infty} \alpha_n \psi_n(x) \quad \text{and} \quad f_2(x) \sim \sum_{n=0}^{\infty} \beta_n \xi_n(x) \quad (x \to x_0) . \quad (2.19)$$

A usual case is when $x_0 = +\infty$, $\gamma(x) = \sin x$, and $\delta(x) = \cos x$. In Chapter 4, we will see that $J_0(x)$ has a compound asymptotic expansion of this type, consisting of Eq. (2.10) together with two asymptotic expansions of the form Eq. (2.19), whose dominant terms appear in Eq. (2.12).

2.5   HISTORICAL AND SUPPLEMENTARY REMARKS

Definitions 2.1– 2.4 are restricted to complex functions of a *real* variable. They are sufficient for the purposes of this book, because our large and small variables are usually real. Extensions to functions of a complex variable can be found in, for example, [2].

The order symbols O and o and the symbol $\sim$ for an asymptotic approximation first appeared in a 1927 book by Bachmann and Landau [2].

Asymptotic expansions have an interesting history, on which it is worth quoting Olver [2]:

"(...) is typical of a large class of divergent series obtained from integral representations, differential equations, and elsewhere when rules governing the applicability of analytical transformations are violated. Nevertheless, such expansions were freely used in numerical and analytical calculations in the eighteenth century by many mathematicians, particularly Euler. (...) little was known about the errors in approximating functions in this way, and sometimes grave inaccuracies resulted. Early in the nineteenth century Abel, Cauchy, and others undertook the task of placing mathematical analysis on firmer foundations. One result was the introduction of a complete ban on the use of divergent series, although it appears that this step was taken somewhat reluctantly.

No way of rehabilitating the use of divergent series was forthcoming during the next half century. Two requirements for a satisfactory general theory were, first, that it apply to most of the known series; secondly, that it permit elementary operations, including addition, multiplication, division, substitution, integration, differentiation, and reversion. Neither requirement would be met if, for example, we confined ourselves to series expansions whose remainder terms are bounded in magnitude by the first neglected term.

Both requirements were satisfied eventually by Poincaré in 1886 by defining what he called asymptotic expansions. (...) Poincaré's theory embraces a wide class of useful divergent series, and the elementary operations can all be carried out (with some slight restrictions in the case of differentiation)."

Generalizations of Definition 2.5 (generalized asymptotic expansions) can be found in [1, 2], and [4].

We have adapted the term compound asymptotic approximation from the term compound asymptotic expansion used in [4].

2.6   PROBLEMS

2.1. For $n = 1, 2, \ldots$, let x_n and y_n stand for the nth positive zero of $\sin \sqrt{x^2 + 1}$ and $\sin x$, respectively. **(i)** Find a simple asymptotic approximation of $x_n - y_n$ as $n \to \infty$, thus demonstrating the increasing closeness of the zeros. **(ii)** Determine the asymptotic power series of $x_n - y_n$ as $n \to \infty$.

2.2. **Addition of asymptotic approximations:**

If $f_1 \sim g_1, f_2 \sim g_2, g_1 \sim \alpha g_2 \ (x \to x_0)$ with $\alpha \neq -1$, show that $f_1 + f_2 \sim g_1 + g_2 \ (x \to x_0)$. Explain why the limitation $\alpha \neq -1$ is necessary.

2.3. **Addition of asymptotic approximations continued:**

If $f_1 = O(g_1), f_2 \sim g_2, g_1 \ll g_2 \ (x \to x_0)$, show that $f_1 + f_2 \sim g_2 \ (x \to x_0)$. ($f_1 = O(g_1)$ can, of course, be replaced by the stronger relation $f_1 \sim g_1$.)

2.4. Use the definitions in Section 2.1 to verify the following asymptotic approximations

$$\ln\left(x + \sqrt{1 + x^2}\right) = x - \frac{x^3}{6} + O\left(x^5\right) \quad (x \to 0^+) \,,$$

$$\ln\left(x + \sqrt{x^2 + 1}\right) \sim \ln\left(2x\right) \quad (x \to +\infty) \,,$$

$$\ln\left(x + \sqrt{x^2 + 1}\right) \sim \ln\left(x\right) \quad (x \to +\infty) \,.$$

Comment on the difference between the second and third relation. Which of the two is better? In what sense?

2.5. Verify that $\exp\left(i\sqrt{4x^3 + 2}\right) \sim \exp\left(i2x^{3/2}\right)$ $(x \to \infty)$. Explain why it is strictly incorrect to add (or subtract) this asymptotic relation to its complex conjugate. This example shows that the real (or imaginary) part of an asymptotic approximation may not itself be an asymptotic approximation—at least in the strict sense of Definition 2.1.

2.6. **Compound asymptotic approximation for Section 2.3.1:**

Show that there exist functions $f(x)$ and $g(x)$ such that

$$\sin\sqrt{x^2 + 1} = f(x)\sin x + g(x)\cos x \,,$$

with

$$f(x) = 1 - \frac{1}{8x^2} + O\left(\frac{1}{x^4}\right) \quad (x \to \infty)$$

and

$$g(x) = \frac{1}{2x} + O\left(\frac{1}{x^3}\right) \quad (x \to \infty) \,.$$

REFERENCES

[1] F. W. J. Olver, D. W. Lozier, R. F. Boisvert, and C. W. Clark, *Digital Library of Mathematical Functions*, National Institute of Standards and Technology from `http://dlmf.nist.gov/`http://dlmf.nist.gov/, §2.1, §10.9.11, §10.25.2, §10.31.1, §10.40.2. 26, 28, 31, 32

[2] F. W. J. Olver, *Asymptotics and Special Functions*. Natick, MA: A. K. Peters, 1997, Chapter 1. 26, 31, 32

[3] C. M. Bender and S. A. Orszag, *Advanced Mathematical Methods for Scientists and Engineers; Asymptotic Methods and Perturbation Theory*. New York: Springer, 1999, §3.7. DOI: 10.1007/978-1-4757-3069-2. 27

[4] R. Wong, *Asymptotic Approximations of Integrals*. Philadelphia: SIAM, 2001. DOI: 10.1137/1.9780898719260. 30, 32

CHAPTER 3

Concepts from Complex Variables

In this chapter, we pause our discussion of asymptotics to deal with some useful topics from the theory of functions of one complex variable. We include (in Section 3.6) some direct applications to antennas and electromagnetics. We assume the reader to already be familiar with concepts such as analytic (regular) functions, isolated singularities, integrals along paths in the complex plane, and the evaluation of integrals using residues and Jordan's lemma.

3.1 GAMMA FUNCTION AND RELATED FUNCTIONS

The **gamma function** $\Gamma(z)$ is one of the most common special functions and plays an important role throughout this book. It is, for example, used extensively in the Mellin-transform method of Chapter 7. Moreover, it appears in the definition of many other special functions. For example, the Bessel function $J_\nu(z)$ can be defined by a power series in z whose ν-dependent coefficients contain gamma functions, see Eq. (A.18) of Appendix A.

For the purposes of this book we define $\Gamma(z)$ as follows,

$$\Gamma(z) = \int_0^\infty t^{z-1}e^{-t}dt, \quad \Re z > 0 , \tag{3.1}$$

which is often called Euler's integral [1]. The constraint $\Re z > 0$ ensures that the integral converges at $t = 0$; see the systematic discussion of such issues in Appendix B. Since $t^{z-1} = e^{(z-1)\ln t}$, the derivative $\Gamma'(z)$ is

$$\Gamma'(z) = \int_0^\infty t^{z-1}e^{-t}\ln t \, dt, \quad \Re z > 0 . \tag{3.2}$$

This integral is convergent by Rule 2 of Appendix B.

Separate the integral in Eq. (3.1) as $\int_0^1 + \int_1^\infty$. In the first integral, expand e^{-t} in a Maclaurin series, interchange the order of summation and integration, and perform the resulting integral to obtain

$$\Gamma(z) = \sum_{n=0}^\infty \frac{(-1)^n}{n!} \frac{1}{z+n} + \int_1^\infty t^{z-1}e^{-t}dt . \tag{3.3}$$

Equation (3.3) holds, initially, for $\Re z > 0$. However, the right-hand side is meaningful for all z except $z \neq 0, -1, -2, \ldots$ and can thus be used to define $\Gamma(z)$ for all such z. Here we "analytically continued" the function originally defined in Eq. (3.1) to an enlarged region; we postpone a systematic discussion of analytic continuation until Section 3.3.

It follows from Eq. (3.3) that $\Gamma(z)$ has simple poles at $z = 0, -1, -2, \ldots$ and that the residues there are

$$\operatorname*{res}_{z=-n} [\Gamma(z)] = \frac{(-1)^n}{n!}, \quad n = 0, 1, 2, \ldots . \tag{3.4}$$

Integrating Eq. (3.1) by parts, we obtain

$$\Gamma(z + 1) = z\Gamma(z) . \tag{3.5}$$

In Section 3.3.4 below, we will invoke analytic continuation to show that Eq. (3.5) is valid not only for $\Re z > 0$, but for all z. Equation (3.5) is the **recurrence formula** for $\Gamma(z)$. By Eq. (3.1), $\Gamma(1) = 1$. From the recurrence formula, it thus follows that

$$\Gamma(n + 1) = n! \quad \text{for} \quad n = 0, 1, 2, \ldots . \tag{3.6}$$

Therefore, $\Gamma(z)$ generalizes the factorial $n!$ (which makes sense only for $n = 0, 1, \ldots$) to complex values of z.

It can be shown (see Problem 3.1 for an outline of a proof) that

$$\Gamma(z)\Gamma(1 - z) = \frac{\pi}{\sin(\pi z)} , \tag{3.7}$$

which is the **reflection formula** [1]. It implies

$$\Gamma\left(\frac{1}{2}\right) = \sqrt{\pi} . \tag{3.8}$$

The **duplication formula** [1] is

$$\Gamma(2z) = \frac{1}{2\sqrt{\pi}} 2^{2z} \Gamma(z) \Gamma\left(z + \frac{1}{2}\right) . \tag{3.9}$$

Equation (3.9) is a special case of **Gauss's multiplication formula** [1]

$$\Gamma(nz) = (2\pi)^{\frac{1-n}{2}} n^{nz - \frac{1}{2}} \prod_{k=0}^{n-1} \Gamma\left(z + \frac{k}{n}\right), \quad n = 1, 2, \ldots . \tag{3.10}$$

For large z, $\Gamma(z)$ possesses a Poincaré-type asymptotic expansion (such expansions were discussed in Section 2.4) of the form [1]

$$\Gamma(z) = \sqrt{2\pi} z^{z-1/2} e^{-z} \left[1 + \frac{1}{12z} + \frac{1}{288z^2} + O\left(\frac{1}{z^3}\right)\right], \quad z \to \infty \quad \text{with} \quad |\mathrm{ph}\, z| < \pi . \tag{3.11}$$

We will derive Eq. (3.11) (or, rather, its special case for z real) in Section 4.2.4 of Chapter 4. Note that Eq. (3.11) does not hold for negative real z. The expansion Eq. (3.11), or sometimes just its leading term, is known as Stirling's formula. Equation (3.11) shows that $\Gamma(z)$ increases very rapidly when z is large and positive, a result consistent with Eq. (3.6).

A plot of $\Gamma(z)$ and $1/\Gamma(z)$ for real z is given in Fig. 3.1. Observe the rapid increase (decrease) of $\Gamma(z)$ ($1/\Gamma(z)$) as z grows. Also observe the poles (zeros) of $\Gamma(z)(1/\Gamma(z))$ at the points $z = 0, -1, -2, \dots$.

The **psi function** $\psi(z)$ is defined by

$$\psi(z) = \frac{\Gamma'(z)}{\Gamma(z)} = \frac{d}{dz} \ln\left[\Gamma(z)\right], \quad z \neq 0, -1, -2, \dots . \tag{3.12}$$

From Eqs. (3.5) and (3.7) we obtain corresponding **recurrence** and **reflection formulas** for $\psi(z)$,

$$\psi(z + 1) = \psi(z) + \frac{1}{z} \tag{3.13}$$

and

$$\psi(1 - z) = \psi(z) + \pi \cot(\pi z) . \tag{3.14}$$

It follows from Eq. (3.13) that

$$\psi(n + 1) = \psi(1) + \sum_{k=1}^{n} \frac{1}{k}, \quad n = 0, 1, 2, \dots . \tag{3.15}$$

The value $\psi(1)$ can be computed as (see Problem 3.2)

$$\psi(1) = -\gamma , \tag{3.16}$$

where $\gamma = 0.57721566\dots$ is **Euler's constant.**

Pochhammer's symbol is

$$(z)_n = \frac{\Gamma(z + n)}{\Gamma(z)}, \quad n = 0, 1, 2, \dots . \tag{3.17}$$

By Eq. (3.5), (3.17) is equivalent to

$$(z)_0 = 1, \quad (z)_n = z(z + 1)(z + 2) \cdot \dots \cdot (z + n - 1), \quad n = 1, 2, \dots , \tag{3.18}$$

and, furthermore, to

$$(z)_0 = 1, \quad (z)_n = (z + n - 1)(z)_{n-1}, \quad n = 1, 2, \dots . \tag{3.19}$$

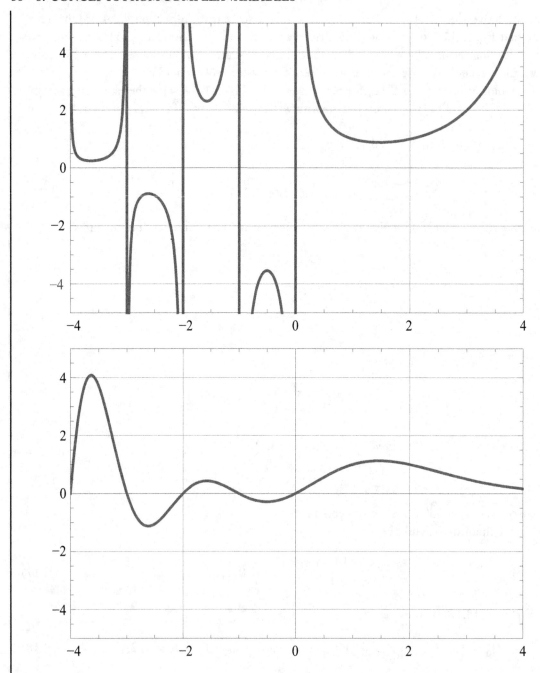

Figure 3.1: $\Gamma(z)$ (top) and $1/\Gamma(z)$ (bottom) as function of z, for real z with $-4 < z < 4$.

3.2 POWER SERIES

In Chapter 2 we showed that power series with a real variable are closely related to asymptotic series and asymptotic approximations. In the present section, we summarize the basics of power series—which are naturally dealt with in the context of complex variables—and give a number of useful examples.

A power series about the point z_0 is a series of the form $\sum_{n=0}^{\infty} \alpha_n(z - z_0)^n$. There exists a ρ with $0 \leq \rho \leq \infty$ such that the series $\sum_{n=0}^{\infty} \alpha_n(z - z_0)^n$ converges for all z with $|z - z_0| < \rho$ and diverges for $|z - z_0| > \rho$. The radius ρ, the circle $|z - z_0| = \rho$, and the disk $|z - z_0| < \rho$ are, respectively, called the **radius, circle,** and **disk of convergence.** The function $f(z) = \sum_{n=0}^{\infty} \alpha_n(z - z_0)^n$ is analytic for all z within the disk of convergence and the power series is called the Taylor series of $f(z)$ about $z = z_0$. When ρ is finite, the circle of convergence passes through the singularity of $f(z)$ that is nearest to z_0 [2]. A Maclaurin series is a Taylor series about $z_0 = 0$.

A function $f(z)$ is analytic at infinity if $f(1/z)$ is analytic at $z = 0$; it thus possesses a convergent series of the form $f(z) = \sum_{n=0}^{\infty} \alpha_n/z^n$ (convergent for $|z|$ greater than some value) which we will call "Taylor series about infinity."

Given a function $f(z)$ analytic at $z = z_0$, the coefficients α_n of its Taylor series about $z = z_0$ are given by

$$\alpha_n = \frac{1}{n!} f^{(n)}(z_0), \quad n = 0, 1, 2, \dots , \tag{3.20}$$

where $f^{(n)}(z) = d^n f(z)/dz^n$ (with $f^{(0)}(z) = f(z)$). The radius of convergence ρ can be found from α_n in a number of ways [3], the simplest of which is the **ratio test;** this states that

$$\rho = \lim_{n \to \infty} \left| \frac{\alpha_n}{\alpha_{n+1}} \right| , \tag{3.21}$$

provided that the limit exists, or equals $+\infty$.

The elementary functions e^z, $\sin z$, $\cos z$, $\sinh z$, and $\cosh z$ are defined through the familiar from real analysis Maclaurin series:

$$e^z = \sum_{n=0}^{\infty} \frac{z^n}{n!}, \quad |z| < \infty , \tag{3.22}$$

$$\sin z = \sum_{n=0}^{\infty} (-1)^n \frac{z^{2n+1}}{(2n+1)!}, \quad |z| < \infty, \tag{3.23}$$

$$\cos z = \sum_{n=0}^{\infty} (-1)^n \frac{z^{2n}}{(2n)!}, \quad |z| < \infty, \tag{3.24}$$

$$\sinh z = \sum_{n=0}^{\infty} \frac{z^{2n+1}}{(2n+1)!}, \quad |z| < \infty, \tag{3.25}$$

$$\cosh z = \sum_{n=0}^{\infty} \frac{z^{2n}}{(2n)!}, \quad |z| < \infty. \tag{3.26}$$

The radii of convergence written in Eqs. (3.22)–(3.26) are found by first dividing Eqs. (3.23) and (3.25) by z, then considering the variable to be z^2 rather than z, and finally applying Eq. (3.21). Let $f(z) = (1+z)^s$, where $s \in \mathbb{C}$. For definiteness, when $s \in \mathbb{C}\backslash\mathbb{Z}$, we mean the principal value of the power function, a concept to be discussed systematically in Section 3.5.2. Then $f^{(n)}(z) = s(s-1)\cdot \ldots \cdot(s-n+1)(1+z)^{s-n}$ for $n = 1, 2, \ldots$. By Eqs. (3.18) and (3.20), the Maclaurin-series coefficients of $f(z)$ are therefore $\alpha_n = (-1)^n(-s)_n/n!$. For $s \neq 0, 1, 2, \ldots$, it follows from Eq. (3.19) that $|\alpha_n/\alpha_{n+1}| = (n+1)/|n-s|$. The ratio test then implies $\rho = 1$, something consistent with the existence of a singularity of $f(z)$ at $z = -1$. We have thus shown the **binomial expansion**

$$(1+z)^s = \sum_{n=0}^{\infty} \frac{(-1)^n(-s)_n}{n!} z^n, \quad |z| < 1. \tag{3.27}$$

When $s = 0, 1, 2, \ldots$ Eq. (3.27) continues to hold but the series terminates and the radius of convergence is infinite (Problem 3.6).

Setting $s = -1$ in Eq. (3.27) and using Eqs. (3.17) and (3.6), we obtain the **geometric series**

$$\frac{1}{1-z} = \sum_{n=0}^{\infty} z^n, \quad |z| < 1, \tag{3.28}$$

which can also be derived from the identity $(1 - z^N)/(1-z) = \sum_{n=0}^{N-1} z^n$. Integrating Eq. (3.28) we obtain

$$\ln(1-z) = -\sum_{n=1}^{\infty} \frac{z^n}{n}, \quad |z| < 1, \tag{3.29}$$

where ln is the principal value of the logarithmic function, to be discussed in Section 3.5.2.

Problem 3.7 describes simple ways to obtain new power series (and, therefore, asymptotic expansions and asymptotic approximations) via the manipulation of known power series. Such manipulations are straightforward, but often messy—it is thus worth mentioning that they can often be carried out by symbolic routines.

3.3   ANALYTIC CONTINUATION

We now move to a systematic discussion of analytic continuation, an important concept with no analogue in functions of a real variable. We begin with some illustrative examples.

3.3.1   REMOVABLE SINGULARITIES

Define $g = \mathbb{C}\backslash\{2,3\}$ and $f(z) = (z-2)/(z^2 - 5z + 6)$. $f(z)$ is analytic throughout g. Since $z^2 - 5z + 6 = (z-2)(z-3)$, we can write $f(z) = 1/(z-3)$. But the right-hand side of this equation is meaningful even when $z = 2$. Furthermore, $1/(z-3)$ is analytic in the extended region $\mathbb{C}\backslash\{3\}$. We have thus analytically continued $f(z)$ to a domain larger than its original domain of definition g.

In this example, the definition was extended in an obvious manner: we simply defined $f(2)$ as a limit, namely as $-1 = \lim_{z \to 2} f(z)$. We say that $z = 2$ is a removable singularity of $f(z)$. Note that $z = 3$ is *not* a removable singularity—at that point, $f(z)$ has a simple pole. Analytic continuation to removable singularities is often done without explicit mention: we automatically think that $\sin z/z$ is equal to 1 when $z = 0$; that $\sinh^2(z-1)/(z-1)$ equals 0 when $z = 1$; that the right-hand side of Eq. (3.5) is meaningful and equal to 1 at $z = 0$; that Eq. (3.5) is valid at $z = -1$ (in the sense, for example, that multiplication by $z + 1$ gives a meaningful equation); and that Eq. (3.17) is equivalent to Eq. (3.18) even when $z = 0, 1, 2, \ldots$, etc.

3.3.2   GEOMETRIC SERIES

Let $f(z) = \sum_{n=0}^{\infty} z^n$. By our previous discussions, $f(z)$ is analytic for $|z| < 1$. By Eq. (3.28), we can think of $(1 - z)^{-1}$ as being an analytic continuation of $f(z)$ from the original domain $|z| < 1$ to the extended domain $\mathbb{C}\backslash\{1\}$. Here, we did not use a limiting process (as in our previous example)—rather, we found an alternative expression for our original function that (i) is meaningful in an extended domain and (ii) defines an analytic function in that extended domain. The analytic function thus defined is the analytic continuation of the original one.

3.3.3   ANALYTIC CONTINUATION DEFINED: UNIQUENESS

Analytic continuation is not always possible. As trivial examples, the function $f(z)$ of Section 3.3.1 cannot be analytically continued to $z = 3$, nor can the $f(z)$ of Section 3.3.2 be analytically continued to $z = 1$. Also, it is non-obvious but true that there are functions defined by power series that cannot be analytically continued beyond the power series's disk of convergence, see [3] and [4] for examples.

If an analytic continuation does exist, is it unique? An answer is given by the following theorem, which we quote (without giving the proof) from [4]; see Fig. 3.2.

Figure 3.2: Regions for Theorem 3.1: Regions G_1 and G_2 have one and only one common region g.

Theorem 3.1   Analytic continuation and uniqueness [4].

   Let a regular function $f_1(z)$ be defined in a region G_1 and let G_2 be another region which has a certain subregion g, but only this one, in common with G_1. Then, if a function $f_2(z)$ exists which is regular in G_2 and coincides with $f_1(z)$ in g, there can only be one such function. $f_1(z)$ and $f_2(z)$ are called analytic continuations of each other. They serve as partial representations or elements of one and the same function $F(z)$ determined by them, and $F(z)$ is regular in the composite region formed by G_1 and G_2.

Remark 3.2   The precise definition of a region is unimportant for our purposes (the interested reader can consult [4]). We only note that a line segment is permissible as a region in Theorem 3.1. Analytic continuation from a segment of the real axis to a larger region of the complex plane is called analytic continuation "into the complex domain." It is unique in the sense of Theorem 3.1.

Remark 3.3   Let us also note that, by the definition in [4], a region is always connected. There is more to be said if g is not connected, meaning that G_1 and G_2 have two (or more) disjoint common parts. We will return to this point shortly.

Remark 3.4   The fact that an analytic function is completely determined (in the sense of Theorem 3.1) by its values in a subdomain—however small this subdomain may be—indicates that the constraint posed by analyticity is a very strong one. In particular, analyticity has no counterpart in functions of a real variable and is a much stronger requirement than the one (in functions of a real variable) of having derivatives of all orders. This point is well-illustrated by the following example: there exist nonzero functions of the real variable x that equal 0 for all $x < 0$ and are everywhere infinitely differentiable; the reader can easily verify that one such function is

$$f(x) = \begin{cases} 0, & x \le 0, \\ e^{-1/x}, & x > 0 \,. \end{cases} \qquad (3.30)$$

On the other hand, the principle of analytic continuation ensures that there is only one *analytic* function of the *complex* variable z that equals 0 when $z \in (-\infty, 0)$—namely, $f(z) = 0$. This marked difference can be traced back to the definition of the derivative, which is much more restrictive in the complex than in the real case.

3.3.4   FURTHER EXAMPLES

These functions in Eqs. (3.22)–(3.29) can now be viewed as *the* analytic continuations of the familiar (from real analysis) functions e^x, $\sin x$, etc., into the complex domain. Equalities such as $\sin(z_1 + z_2) = \sin z_1 \cos z_2 + \cos z_1 \sin z_2$, known to hold for real z_1 and z_2, must—by the principle of analytic continuation—also hold when z_1 and z_2 are complex.

Equation (3.5) was derived from Eq. (3.1), an equation meaningful only for $\Re z > 0$. Nonetheless, by analytic continuation Eq. (3.5) must hold for all z. Even the points $z = 0, -1, -2, \ldots$ need not be excluded: the equation can be rearranged to give $\Gamma(z + 1)/\Gamma(z) = z$, which is always meaningful (at $z = 0, -1, -2, \ldots$, the left-hand side is understood, as usual, as a limiting value).

3.3.5   INTEGRALS THAT ARE ANALYTIC FUNCTIONS OF A PARAMETER

The following two theorems, quoted from [3], give sufficient conditions for integrals of the form $\int_C f(t, z) dt$ to be analytic functions of z.

Theorem 3.5   Analytic functions represented by finite integrals [3].
   Let $f(t, z)$ satisfy the following conditions when t lies on a certain [finite] path of integration (a, b) and z is any point of a region S:
   (i) f and $\frac{\partial f}{\partial z}$ are continuous functions of t.
   (ii) f is an analytic function of z.
   (iii) The continuity of $\frac{\partial f}{\partial z}$ qua function of z is uniform with respect to the variable t.
   Then $\int_a^b f(t, z)\, dt$ is an analytic function of z [and] has the unique derivative $\int_a^b \frac{\partial f(t,z)}{\partial z}\, dt$.

Theorem 3.6   Analytic functions represented by infinite integrals [3].

$\int_a^\infty f(t, z)\, dt$ is an analytic function of z at all points of a region S if (i) the integral converges, (ii) $f(t, z)$ is an analytic function of z when t is on the path of integration and z is on S, (iii) $\frac{\partial f(t,z)}{\partial z}$ is a continuous function of both variables, (iv) $\int_a^\infty \frac{\partial f(t,z)}{\partial z}\, dt$ converges uniformly throughout S.

For if these conditions are satisfied, $\int_a^\infty f(t, z)\, dt$ has the unique derivative $\int_a^\infty \frac{\partial f(t,z)}{\partial z}\, dt$.

Thus, if the conditions of Theorem 3.5 (or, for the case of an infinite integration limit, those of Theorem 3.6) are satisfied, the function of z defined by the integral $\int_a^b f(t, z)\, dt$ (or $\int_a^\infty f(t, z)\, dt$) is analytic, and its derivative is found by differentiating under the integral sign.

By Theorem 3.6, the integral in Eq. (3.1) is analytic when $\Re z > 0$, with derivative given by Eq. (3.2). Similarly, the integral in Eq. (3.3) is analytic when $z \neq 0, -1, -2, \ldots$. A further example involving a special function is given in Problem 3.10.

3.4   MULTIVALUED FUNCTIONS AND BRANCH POINTS

Now consider regions G_1 and G_2 with *two* common subregions g and h, as shown in Fig. 3.3(a). If $f_1(z)$ and $f_2(z)$ are defined in G_1 and G_2, respectively, and if $f_1(z) = f_2(z)$ in g, then is it necessarily true that $f_1(z) = f_2(z)$ in h? Note that Theorem 3.1 does not provide an answer to this question.

An equivalent question is: We wish to analytically continue a function originally defined in g to a new region h (with h disjoint from g, as in Fig. 3.3(a)). If we perform this analytic continuation along two different regions G_1 and G_2, we obtain two functions defined in h. Are the two functions necessarily equal?

Yet another related question is: We analytically continue a function, originally defined in g, along a region G, like the one shown in Fig. 3.3(b): G is not simply connected and is such that we end up in g again. Thus, we finally obtain another function defined in g. Are the original and final functions necessarily equal in g?

The examples below demonstrate that the answer to all three questions is no. In light of Theorem 3.1 this is, perhaps, surprising, but we have already stressed that Theorem 3.1 does not cover the situations we are now discussing.

3.4.1   SQUARE ROOT

In our first example, we specifically address the third of our three questions: We wish to analytically continue the familiar from real variables square root, i.e., the positive function $f_1(z) = \sqrt{z}$ (z real and positive), end up where we started (i.e., at the positive real axis), and examine whether the final function is equal to the original one. Let $z = re^{i\theta}$, where $r > 0$. Analytic continuation

Figure 3.3: (a) Regions G_1 and G_2 with two common parts, g and h. (b) Region G that is not simply connected and its subregion g.

can be done along any region G by means of the formula

$$f(z) = \sqrt{r}e^{i\theta/2} \tag{3.31}$$

(where $\sqrt{r} > 0$, for definiteness), provided that (i) G does not include the origin; (ii) the initial value of θ is $\theta = 0$ (it is also possible to take $\theta = 4\pi, 8\pi, \ldots$, but not $\theta = 2\pi, 6\pi, \ldots$); and (iii) θ is continuous throughout G: if (i)–(iii) hold, $f(z)$ initially coincides with $f_1(z)$, and $f(z)$ is clearly analytic throughout G ($f(z)$ is not analytic at $z = 0$, a point we explicitly excluded). It is convenient to think of starting at some point A on the positive real semi-axis, analytically continuing along a closed path, and ending up at A again, as in Figs. 3.4(a) and 3.4(b).

By tracing values of θ for the two paths drawn in Fig. 3.4(a), the final value of θ is seen to be the same as the starting value, i.e., $\theta = 0$; in this case, therefore, the final value of $f(z)$ equals the initial value (and is positive). But for the path in Fig. 3.4(b), the final value of θ is $\theta = 2\pi$ rather than $\theta = 0$; in such a case, the final value of $f(z)$—as given by Eq. (3.31)—does not equal the initial value (the final value is now negative).

Readers should convince themselves that the distinguishing feature is this: the paths of Fig. 3.4(a) do not encircle the origin, while that of Fig. 3.4(b) encircles it exactly once. A single encirclement of $z = 0$ always causes a change in value, however small the encirclement may be. And the origin $z = 0$ is the only point (the only finite point, that is; the point at infinity will be discussed shortly) possessing the above-described property. For these reasons, $z = 0$ is called a **branch point** of $\sqrt{z}$. Branch points are associated with multivalued functions, with our example

Figure 3.4: (a) Closed paths that start at a point A on the positive real semi-axis and do not encircle the branch point at the origin. (b) Closed path that starts at point A and encircles the origin exactly once.

$\sqrt{z}$ being two-valued. Thinking in the same manner, we find that $z = 0$ is the only finite branch point of the n-valued function $z^{1/n}$, $n = 2, 3, \ldots$ (see also Problem 3.12).

3.4.2 FURTHER EXAMPLES

In Table 3.1, we have written the finite branch points of some functions that we consider particularly instructive. We recommend readers to persuade themselves of the correctness of Table 3.1, aided, if necessary, by the following remarks: In Row 2, we follow [1] in denoting the multivalued or general logarithmic function by Ln z. This function can be understood both through its definition by means of an integral (written in Table 3.1) and, also, through the relation

$$\text{Ln } z = \ln r + i\theta \quad (z = re^{i\theta}) \tag{3.32}$$

(where $\ln r$ real, for definiteness). With the integral definition, multivaluedness arises because of the multiplicity of choices for the integration path (see also Section 3.5.2 below); with Eq. (3.32), multivaluedness arises because of the infinite number of choices for θ, with the various choices differing by integer multiples of 2π. As $\cos z$ is an even function, $\cos \sqrt{z}$ (Entry 5) is not multivalued and has no branch points. An analogous statement holds for Entry 7: $z = 0$ is a removable singularity of $\sin \sqrt{z}/\sqrt{z}$, whose Maclaurin series follows immediately from Eq. (3.23). The function in Entry 9 is two-valued but there is no branch point because analytic continuation around any point (and, in particular, around $z = 0$) will leave the function unchanged.

Table 3.1: Some functions and their finite branch points

3.4.3 THE POINT AT INFINITY

The point at infinity is a branch point for $f(z)$ if $w = 0$ is a branch point for $g(w) = f(1/w)$. The idea is illustrated by the examples of Table 3.2. Note that the last column contains equalities between multivalued functions, with the functions in the equalities' right-hand sides attaining precisely the same values (no more and no less) as those on the left, e.g., $(1 - w^2)^{1/2}/w$ attains two values, just like $\left[(1/w)^2 - 1\right]^{1/2}$ —these values are identical and the two-valued functions are equal.

If $z = \infty$ is a branch point, then analytic continuation of a function along every closed and sufficiently *large* path results in a different final function upon return.

Table 3.2: Functions with/without $z = \infty$ as a branch point

3.5   BRANCHES AND PRINCIPAL VALUES OF MULTIVALUED FUNCTIONS

From a multivalued function, define a one-valued function by assigning one of the possible function values to each point z in some region where the function is defined. If the one-valued function is analytic throughout the region, it is called a **branch** of the multivalued function. It is not always possible to construct a branch within a given region. Since $z = 0$ is a branch point of both $\sqrt{z}$ and of $\ln z$, for example, neither of these functions has a branch defined in the punctured region $\mathbb{C}\backslash\{0\}$. As we now proceed to illustrate, however, we can construct branches of these function in regions that are "cut" in a proper manner. In what follows, $z = re^{i\theta}$.

3.5.1   SQUARE ROOT

First consider the two-valued function $\sqrt{z}$. In the cut region $\mathbb{C}\backslash(-\infty, 0]$ (i.e., the entire complex plane with the exception of the nonpositive real semi-axis, see Fig. 3.5(a)), we can define the following two branches:

$$\text{first or principal branch:} \quad \sqrt{z} = \sqrt{r}e^{i\theta/2}, \quad -\pi < \theta < \pi, \tag{3.33}$$

$$\text{second branch:} \quad \sqrt{z}\big|_2 = \sqrt{r}e^{i\theta/2}, \quad \pi < \theta < 3\pi. \tag{3.34}$$

Figure 3.5: Cut regions for the definition of branches for $\sqrt{z}$, Ln z, and z^α ($\alpha \notin \mathbb{Z}$). Branch cuts are shown with dashed lines.

The value given by the principal branch is the **principal value** of the square root. It is positive when z is real and positive. It is usual to denote the principal branch by $\sqrt{z}$ (instead of, say, $\sqrt{z}|_1$, a more awkward notation that would risk less confusion with the two-valued function $\sqrt{z}$ or with other branches). The cut at $(-\infty, 0]$, which prevents a complete traversal about the branch point $z = 0$, is called a **branch cut** for $\sqrt{z}$. This cut extends to infinity, consistent with the existence of a branch point at $z = \infty$. Note that the two branches can be specified in many equivalent ways. For example, the principal branch can also be designated as $\sqrt{r}e^{i\theta/2}$, $\quad 3\pi < \theta < 5\pi$, and the branch in Eq. (3.34) is the same as $-\sqrt{r}e^{i\theta/2}$, $\quad -\pi < \theta < \pi$ (so that, for any z in the cut plane, the value of this branch is the opposite of the principal value).

The two branches in Eq. (3.33) and Eq. (3.34) are connected in the sense that each branch is the analytic continuation of the other one across the branch cut. Thus, the principal value right above the point $z = -1$ equals the value of the second branch right below -1: $\sqrt{-1 + i0} = \sqrt{-1 - i0}|_2 = i$. Similarly, $\sqrt{-1 - i0} = \sqrt{-1 + i0}|_2 = -i$.

Another possible cut region is shown in Fig. 3.5(b). Here, the branch cut is $[0, +\infty)$. The two possible branches consistent with this cut are

$$\text{first branch:} \qquad \sqrt{z}\big|_1 = \sqrt{r}e^{i\theta/2}, \quad 0 < \theta < 2\pi, \tag{3.35}$$
$$\text{second branch:} \qquad \sqrt{z}\big|_2 = \sqrt{r}e^{i\theta/2}, \quad 2\pi < \theta < 4\pi. \tag{3.36}$$

Neither of these branches is customarily referred to as a principal branch.

We stress the freedom in choosing branch cuts—usually, the choice is dictated by the requirements of the particular problem.

3.5.2   LOGARITHM AND POWERS OTHER THAN THE SQUARE ROOT

The cut regions of Figs. 3.5(a) and 3.5(b) are also the most usual ones for the definition of branches of Ln z. For the cut region of Fig. 3.5(a), the possible branches are

$$\text{Ln } z\big|_m = \ln r + i(\theta + 2m\pi), \quad -\pi < \theta < \pi, \quad m \in \mathbb{Z}. \tag{3.37}$$

Here there are infinitely many branches, each associated with an integer m. Branch m is connected to two other branches, namely $m + 1$ and $m - 1$. For the cut region of Fig. 3.5(b), the branches are

$$\text{Ln } z\big|_m = \ln r + i(\theta + 2m\pi), \quad 0 < \theta < 2\pi, \quad m \in \mathbb{Z}. \tag{3.38}$$

Although Eqs. (3.37) and (3.38) are associated with different branch cuts, we have used the same symbol $\text{Ln } z\big|_m$; this should be avoided if there is a risk of confusion.

The **principal branch** is the branch in Eq. (3.37) with $m = 0$. It is usually denoted by $\ln z$, so that

$$\ln z = \ln r + i\theta, \quad -\pi < \theta < \pi. \tag{3.39}$$

This branch reduces to the familiar logarithm when z is real and positive. An equivalent definition is

$$\ln z = \int_1^z \frac{dt}{t}, \quad z \notin (-\infty, 0], \tag{3.40}$$

where the path of integration does not intersect $(-\infty, 0]$, cf. the integral definition of Ln z in Table 3.1. By Cauchy's integral formula, the other branches in Eq. (3.37) can be defined similarly, but with different integration paths. For example, for any value of z in the upper-half complex t-plane, a proper integration path for the branch $\text{Ln } z\big|_1$ of Eq. (3.37) is shown in Fig. 3.6.

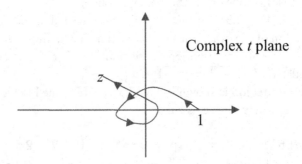

Complex t plane

Figure 3.6: Path of integration for the branch $\text{Ln } z\big|_1$ of (3.40), for $\Im z > 0$.

The function z^α can be expressed in terms of the exponential function and the multivalued logarithm $\text{Ln}\, z$ using

$$z^\alpha = e^{\alpha\, \text{Ln}\, z} \,. \tag{3.41}$$

Except when $\alpha \in \mathbb{Z}$, z^α is multivalued with branch points at $z = 0$ and $z = \infty$. Its principal branch is defined in the cut plane of Fig. 3.5(a) and involves the principle branch of the logarithm:

$$z^\alpha = e^{\alpha\, \ln z} = e^{\alpha\, (\ln r + i\theta)}, \quad -\pi < \theta < \pi \,. \tag{3.42}$$

3.5.3   THE FUNCTION $\sqrt{z^2 - 1}$

Recall from Tables 3.1 and 3.2 that $\sqrt{z^2 - 1} = \sqrt{(z + 1)(z - 1)}$ is two-valued, with branch points at $z = -1$ and $z = 1$, and with no branch point at infinity. Two possible choices of cut planes are shown in Figs. 3.7 and 3.8. The configuration in Fig. 3.8 is acceptable because there is

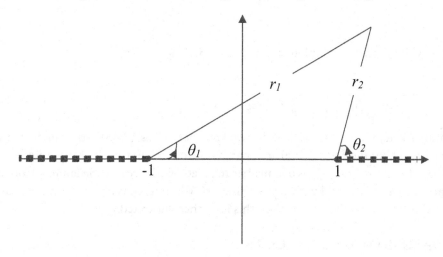

Figure 3.7: Cut region for definition of branches in (3.43) and (3.44).

no branch point at infinity. Explicit expressions for the branches can be found by setting $z + 1 = r_1 e^{i\theta_1}$ and $z - 1 = r_2 e^{i\theta_2}$, and specifying the ranges of θ_1 and θ_2. Thus, there are two branches in the cut plane of Fig. 3.7,

$$\sqrt{z^2 - 1}\Big|_1 = \sqrt{r_1 r_2}\, e^{i(\theta_1 + \theta_2)/2}, \quad -\pi < \theta_1 < \pi, \quad 0 < \theta_2 < 2\pi \tag{3.43}$$

and

$$\sqrt{z^2 - 1}\Big|_2 = \sqrt{r_1 r_2}\, e^{i(\theta_1 + \theta_2)/2}, \quad -\pi < \theta_1 < \pi, \quad 2\pi < \theta_2 < 4\pi \,. \tag{3.44}$$

The two branches consistent with Fig. 3.8 are

$$\sqrt{z^2 - 1}\Big|_1 = \sqrt{r_1 r_2}\, e^{i(\theta_1 + \theta_2)/2}, \quad 0 < \theta_1 < 2\pi, \quad 0 < \theta_2 < 2\pi \tag{3.45}$$

Figure 3.8: Cut region for definition of branches in (3.45) and (3.46).

and

$$\left.\sqrt{z^2 - 1}\right|_2 = \sqrt{r_1 r_2}\, e^{i(\theta_1 + \theta_2)/2}, \quad 2\pi < \theta_1 < 4\pi, \quad 0 < \theta_2 < 2\pi . \tag{3.46}$$

The reader is asked to check that each of the two branches in Eqs. (3.43) and (3.44) is an even function of z (throughout the cut z-plane) and that, on the other hand, each of Eqs. (3.45) and (3.46) is odd. That a branch of $\sqrt{z^2 - 1}$ is odd may seem paradoxical, because the familiar from real variables function $\sqrt{z^2 - 1}$ (z real with $|z| > 1$) is even. We must thus be careful when dealing with multivalued functions and their branches; this is further illustrated in Problems 3.14 and 3.15.

3.5.4   VALUES ON BRANCH CUTS

None of our branches are defined *on* their respective branch cuts and it is often useful to do so. For example, the principal value of the logarithm is often defined to be [1]

$$\ln z = \ln r + i\theta, \quad -\pi < \theta \le \pi . \tag{3.47}$$

As another example, define θ_1 and θ_2 as in Fig. 3.7. The following branch of $\sqrt{z^2 - 1}$,

$$\left.\sqrt{z^2 - 1}\right|_1 = \sqrt{r_1 r_2}\, e^{i(\theta_1 + \theta_2)/2}, \quad -\pi \le \theta_1 < \pi, \quad 0 \le \theta_2 < 2\pi , \tag{3.48}$$

is a minor modification of Eq. (3.43) that reduces to the familiar from real variables function $\sqrt{z^2 - 1}$ when z is real with $|z| > 1$. It is one-valued and even throughout $\mathbb{C}\backslash\{-1, 1\}$.

Also in use is the so-called **closed definition** of the principal value of Ln z [1]. This is defined by Eq. (3.39) together with

$$\ln(x \pm i0) = \ln|x| \pm i\pi, \quad x \in (-\infty, 0) , \tag{3.49}$$

where $x \pm i0$ denotes the limiting value from the upper (lower) side of the cut. With this definition, $\ln(1/z) = -\ln z$ for all nonzero z. Such closed definitions are useful when paths of integration run along branch cuts. For examples, see Problems 3.17 to 3.19.

3.6   APPLICATIONS TO ANTENNAS AND ELECTROMAGNETICS: NONSOLVABILITY

In this section, we present some direct applications of the concepts developed in Sections 3.1–3.5 to antennas and computational electromagnetics. We specifically use analytic continuation arguments to demonstrate that certain integral equations relevant to wire antennas and to the Method of Auxiliary Sources (MAS) have no solutions.

3.6.1   HALLÉN'S AND POCKLINGTON'S EQUATIONS WITH THE APPROXIMATE KERNEL

Hallén's integral equation for the current $I(z)$ on a linear antenna is discussed in most antenna textbooks. For the case where the antenna is center-driven by a delta-function generator and with a $e^{-i\omega t}$ time dependence, this equation is [5, 6]

$$\int_{-h}^{h} K(z - z')I(z')\,dz' = \frac{iV}{2\zeta_0}\sin k\,|z| + C\cos kz, \quad -h < z < h, \tag{3.50}$$

where $2h$ is the length of the antenna, $k = \omega/c$ is the free-space wavenumber, and $\zeta_0 = 376.73$ Ohms is the free-space impedance. V is the driving voltage at $z = 0$, and C is a constant to be determined from the condition $I(\pm h) = 0$. The so-called approximate kernel is [5]

$$K(z) = \frac{1}{4\pi}\frac{\exp\left(ik\sqrt{z^2 + a^2}\right)}{\sqrt{z^2 + a^2}}, \tag{3.51}$$

where a is the radius, assumed to satisfy $ka \ll 1$ and $a \ll h$. It was known as far back as 1952 [7] that, with Eqs. (3.51), (3.50) is nonsolvable. In other words, no function $I(z)$ can satisfy Eq. (3.50). This important fact—frequently not mentioned in modern textbooks—is fairly obvious: Under mild admissibility conditions on $I(z)$, the left-hand side of Eq. (3.50) is differentiable (with differentiation permissible under the integral sign). On the other hand, the right-hand side of Eq. (3.50) is not differentiable at $z = 0$ because of the absolute value. Thus, $I(z)$ cannot satisfy Eq. (3.50). Physically, the approximate integral equation requires a line current located on the z-axis to maintain a field with an abrupt behavior at the location $(\rho, z) = (a, 0)$ of the delta-function generator, and this is not possible.

Here, we follow [5] and [6] and extend this result to: (i) the case of Hallén's equation with plane-wave incidence, and (ii) the case of Pocklington's equation for an antenna driven by the "magnetic frill generator" described in [8].

For the case of plane-wave incidence, Hallén's equation is [5]

$$\int_{-h}^{h} K(z - z')I(z')\,dz' = \frac{iV}{\zeta_0} + C\cos kz, \quad -h < z < h, \tag{3.52}$$

where V is a constant with dimension of Volts. As opposed to the right-hand side of Eq. (3.50), that of Eq. (3.52) is differentiable so that the nonsolvability argument used for Eq. (3.50) is not applicable. Let us suppose that Eq. (3.52) has a solution that is continuous,

$$I(z) = \text{continuous} \quad (-h < z < h), \tag{3.53}$$

an assumption that seems adequate for all practical purposes. Now think of z as a complex variable and consider values of z not lying on the line segment $(-h, h)$. By Eq. (3.51) and Theorem 3.5, the left-hand side of Eq. (3.52) is an analytic function of the complex variable z with the *possible* exception of points z that satisfy $z' - z = \pm ia$ for some $z' \in [-h, h]$. These points form two line segments centered at $z = \pm ia$ and parallel to the real axis; each segment has length $2h$.[1] Now analytically continue the left-hand side of Eq. (3.52) to large, positive values of z along any path not intersecting the aforementioned two segments. By Eq. (3.51), one thus obtains a function decaying like $\exp(ikz)/z$ for large z. For any choice of C, however, the right-hand side of Eq. (3.52) cannot behave like $\exp(ikz)/z$ for large z. Hence, the two functions defined for $z \in (-h, h)$ have analytic continuations which cannot be the same. Therefore, the functions themselves cannot be the same. We have thus reached a contradiction showing that Eq. (3.52) cannot have a solution satisfying Eq. (3.53).

For the case of the magnetic frill generator, Pocklington's equation is [6]

$$\left(\frac{\partial^2}{\partial z^2} + k^2\right) \int_{-h}^{h} K(z - z')I(z')\,dz' = g(z), \quad -h < z < h, \tag{3.54}$$

where

$$g(z) = \frac{ikV}{2\zeta_0 \ln(b/a)} \left[\frac{\exp\left(ik\sqrt{z^2 + a^2}\right)}{\sqrt{z^2 + a^2}} - \frac{\exp\left(ik\sqrt{z^2 + b^2}\right)}{\sqrt{z^2 + b^2}} \right]. \tag{3.55}$$

Here, b is the outer radius of the frill and V is the equivalent to the frill driving voltage. Assuming a solution that satisfies Eq. (3.53) and arguing as before, we see that the left-hand side of Eq. (3.54) is analytic for all $z \in \mathbb{C}$ with the possible exception of the two aforementioned line segments. The right-hand side, however, has branch points at $z = \pm ib$ (it also has branch points at $z = \pm ia$, not of interest herein). Since $b > a$, these two points cannot lie on the aforementioned line segment

---

[1]The segments need not be branch cuts. To gain insight into problems like this, it is instructive to work out the simple integral $\int_{-h}^{h}[(z - z')^2 + a^2]^{-1/2}dz'$.

and, once again, we have a contradiction showing that Eq. (3.54) cannot have a solution satisfying Eq. (3.53). We note that Hallén's equation cannot have such a solution, either; this is shown via a different analytic continuation argument in [6].

We note that the above arguments are not applicable for the case where the antenna length is infinite, $h = \infty$ [6]. This fictitious antenna is in many respects simpler than its finite counterpart and can often help one understand the actual finite antenna; we will encounter the infinite antenna several times in this book.

3.6.2   INTEGRAL EQUATION RELATED TO THE METHOD OF AUXILIARY SOURCES (MAS)

The Method of Auxiliary Sources (MAS) is an approximate method for the solution of electromagnetic scattering problems. An overview of MAS is given in [10]. In the simplest case, that of scattering by a closed, smooth, perfect electric conductor (PEC) illuminated externally, one assumes electric currents—to be referred to here as MAS currents—on N fictitious sources located on an auxiliary surface inside the PEC. The MAS currents are such that the boundary condition of vanishing tangential electric field on N collocation points on the PEC scatterer is satisfied. A $N \times N$ system of linear algebraic equation thus results. Once the MAS currents are found, the field due to them—to be called MAS field—can be easily determined. As N grows, one hopes for convergence of the MAS field to the true scattered field. Furthermore, when the auxiliary surface is smooth and closed (this is usually the case in the literature), it is natural to anticipate that—when properly normalized—the MAS currents should remain unchanged. As N grows, that is, one hopes for a better approximation to the field with only small changes in the corresponding normalized MAS currents. In other words, one hopes that the normalized MAS currents converge to corresponding surface current densities.

In [9], it is shown that it is possible for the MAS field to converge to the true scattered field without having the normalized MAS currents converge. This is accomplished through a study of a simple two-dimensional problem, in which the scatterer is an infinitely long, PEC circular cylinder illuminated by an infinitely long, constant-current line filament. The MAS currents are equispaced and lie on a circle, and so do the N collocation points. For this simple case, the matrix in the $N \times N$ system of algebraic equations is circulant and an explicit, closed-form solution for the MAS currents and fields can be obtained. The main conclusions in [9] then follow by developing asymptotic approximations for currents and fields as $N \to \infty$.

In [9] much understanding is obtained via the "continuous version of MAS," where a continuous surface current density $J^S(\phi_{aux})$, located on the auxiliary surface, is determined from an integral equation arising from the exact satisfaction of the relevant boundary condition. As one might expect, the normalized MAS currents converge to the solution of this integral equation in the case where the equation is solvable; they diverge, however, when the equation is nonsolvable. Here, we present an analytic continuation argument [9] that gives a necessary condition for nonsolvability.

With an $e^{-i\omega t}$ time dependence, the integral equation for the unknown auxiliary surface current density $J^S(\phi_{aux})$ is [9]

$$\int_{-\pi}^{\pi} H_0^{(1)}(kR_{cyl,aux})J^S(\phi_{aux})d\phi_{aux} = -\frac{I}{\rho_{aux}}H_0^{(1)}(kR_{cyl,fil}), \quad -\pi < \phi_{cyl} < \pi. \quad (3.56)$$

Here, k is the wavenumber, $H_0^{(1)}(kR_{cyl,aux})$ is the kernel of the integral equation—where $H_0^{(1)}$ is the first Hankel function (see Appendix A)—and $R_{cyl,aux}$ is the distance from a point (ρ_{aux}, ϕ_{aux}) on the auxiliary circular surface to a point (ρ_{cyl}, ϕ_{cyl}) on the PEC circular cylinder, see Fig. 3.9. Thus,

$$R_{cyl,aux} = \sqrt{\rho_{cyl}^2 + \rho_{aux}^2 - 2\rho_{cyl}\rho_{aux}\cos(\phi_{cyl} - \phi_{aux})}. \quad (3.57)$$

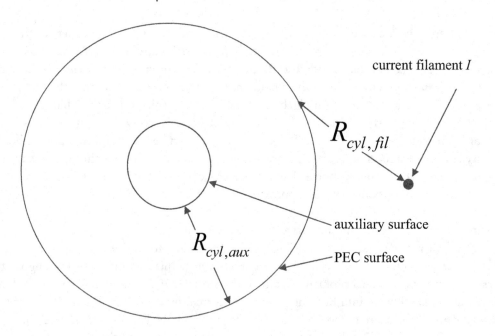

Figure 3.9: Geometry of simple scattering problem and distances appearing in (3.56).

Similarly, $R_{cyl,fil}$ is the distance from (ρ_{cyl}, ϕ_{cyl}) to $(\rho_{fil}, 0)$, where $\rho_{cyl} < \rho_{fil}$ and $(\rho_{fil}, 0)$ is the point where the illuminating source (current filament I) is located:

$$R_{cyl,fil} = \sqrt{\rho_{cyl}^2 + \rho_{fil}^2 - 2\rho_{cyl}\rho_{fil}\cos\phi_{cyl}}. \quad (3.58)$$

Equation (3.56) states that $J^S(\varphi_{aux})$ is such that the field produced by it (and found via the superposition integral on the left-hand side of Eq. (3.56)) cancels the incident field (right-hand side of Eq. (3.56)) at all points (ρ_{cyl}, ϕ_{cyl}) on the PEC surface.

We now show that Eq. (3.58) has no solution satisfying

$$J^S(\phi_{aux}) = \text{continuous} \quad (-\pi < \phi_{aux} \leq \pi) \tag{3.59}$$

in the following case

$$\rho_{aux} < \rho_{cri} \equiv \frac{\rho_{cyl}^2}{\rho_{fil}}, \tag{3.60}$$

i.e., when the radius of the auxiliary surface is smaller than the "critical radius" ρ_{cri} defined in Eq. (3.60). Note that $\rho_{cyl} < \rho_{fil}$ implies $\rho_{cri} < \rho_{cyl}$.

Reasoning as in Section 3.6.1, we think of ϕ_{cyl} as a complex variable. By Appendix A, $H_0^{(1)}(z)$ has a branch point when $z = 0$. Therefore Eq. (3.58) shows that the right-hand side of Eq. (3.56) has branch points when

$$\cos\phi_{cyl} = \frac{\rho_{cyl}^2 + \rho_{fil}^2}{2\rho_{cyl}\rho_{fil}} = \frac{\rho_{cri}^2 + \rho_{cyl}^2}{2\rho_{cri}\rho_{cyl}}, \tag{3.61}$$

where the second expression follows from Eq. (3.60). The complex solutions ϕ_{cyl} to Eq. (3.61)—whose right-hand side is greater than 1—can be found via the identity

$$\cos(x + iy) = \cos x \cosh y - i \sin x \sinh y. \tag{3.62}$$

By Eq. (3.62), (3.61) is satisfied when $\phi_{cyl} = \psi^{(n,+)}$ and $\phi_{cyl} = \psi^{(n,-)}$, where

$$\psi^{(n,\pm)} = 2n\pi \pm i \operatorname{arccosh}\frac{\rho_{cri}^2 + \rho_{cyl}^2}{2\rho_{cri}\rho_{cyl}}, \quad n \in \mathbb{Z}, \tag{3.63}$$

where we use the usual symbol [1] for the real-valued inverse hyperbolic function (in other words, the principal value of the multivalued function Arccosh). The branch points $\psi^{(n,\pm)}$ are shown in Fig. 3.10.

By Eq. (3.59) and Theorem 3.5, the left-hand side of Eq. (3.56) is an analytic function of the complex variable ϕ_{cyl}, with the possible exception of points ϕ_{cyl} that are singularities of the integrand for some $\phi_{aux} \in [-\pi, \pi]$. Proceeding from Eq. (3.57) and arguing as before, we find that these singularities are the points $\phi_{cyl} = \xi^{(n,+)}$ and $\phi_{cyl} = \xi^{(n,-)}$ where

$$\xi^{(n,\pm)} = \phi_{aux} + 2n\pi \pm i \operatorname{arccosh}\frac{\rho_{aux}^2 + \rho_{cyl}^2}{2\rho_{aux}\rho_{cyl}}, \quad n \in \mathbb{Z} \quad \phi_{aux} \in [-\pi, \pi]. \tag{3.64}$$

As ϕ_{aux} varies, the points in Eq. (3.64) form two horizontal lines. It follows that the left-hand side of (3.56) is analytic within the strip between these lines. If the inequality Eq. (3.60) is satisfied, we will show that the points in Eq. (3.64) are such that

$$|\operatorname{Im}\{\psi^{(n,\pm)}\}| < |\operatorname{Im}\{\xi^{(n,\pm)}\}|, \tag{3.65}$$

Figure 3.10: When (3.60) is satisfied, the singularities of the right-hand side (RHS) of (3.56) lie within the strip of analyticity of the left-hand side. Thus, (3.60) cannot have a solution. *(Figure from Fikioris [9]).*

as shown in Fig. 3.10. Thus, the branch points of the right-hand side fall within the analyticity strip of the left-hand side. We thus have reached a contradiction showing that, subject to Eq. (3.60), the integral equation Eq. (3.56) cannot have a solution satisfying Eq. (3.59).

We stress that our nonsolvability argument is only valid subject to Eq. (3.60). Equation (3.65) is not satisfied when $\rho_{cri} < \rho_{aux} < \rho_{cyl}$, which means that the auxiliary radius is larger than the critical radius. In that case, Eq. (3.56) does have a solution, but this must be shown by other means. (Due to the simplicity of the problem, the solution can in fact be determined in closed form [9].)

We close this section by showing Eq. (3.65): Since $\cosh x$ is an increasing function for $x > 0$, $\operatorname{arccosh} x$ is also an increasing function, and it suffices to show that

$$\frac{\rho_{cri}^2 + \rho_{cyl}^2}{2\rho_{cri}\rho_{cyl}} < \frac{\rho_{aux}^2 + \rho_{cyl}^2}{2\rho_{aux}\rho_{cyl}} . \tag{3.66}$$

Rearrangement gives the equivalent inequality $(\rho_{aux} - \rho_{cri})(\rho_{cyl}^2 - \rho_{aux}\rho_{cri}) < 0$, which is true because $0 < \rho_{aux} < \rho_{cri} < \rho_{cyl}$.

3.7   SUPPLEMENTARY REMARKS AND FURTHER READING

The material in Sections 3.1– 3.5 is developed in textbooks on the theory of one complex variable. We have already referred to and quoted from the classic works by Whittaker and Watson [3] and Knopp [4]. There are many more excellent pertinent textbooks, e.g., [11]–[17]. We can also recommend the concise coverage of topics relevant to complex variables in the NIST *Digital Library of Mathematical Functions* [1].

Besides Euler's definition Eq. (3.1) of the gamma function, there exist many others; two are given in Problems 3.1 and 3.19. Most properties of $\Gamma(z)$ were developed by eminent 19th century mathematicians. The study of these properties—in addition to being useful in their own right—is excellent practice in functions of a complex variable. For more properties, detailed proofs, and historical notes, the interested reader can consult [3]. A historical profile of $\Gamma(z)$ from its invention until the mid-20th century can be found in [18].

For completeness, let us here give the actual definition of the constant γ, also called the Euler-Mascheroni constant. It is [1]

$$\gamma = \lim_{n\to\infty} \left(\frac{1}{1} + \frac{1}{2} + \ldots + \frac{1}{n} - \ln n \right) . \tag{3.67}$$

There are also series and integral representations for γ, see Problem 3.2 for a sample. Euler estimated γ to a precision of 16 decimal digits. In 1999, γ was calculated to 108,000,000 decimal digits. Today, much is known about γ and there is an entire book devoted to it [19]. Yet we still do not know if γ is an irrational number. The famous British mathematician G. H. Hardy (1877–1947) offered to vacate his Savilian Chair at Oxford to anyone who could prove that γ is irrational [19].

Our notation for Pochhammer's symbol agrees with that of [1] and is widespread in the literature of special functions. We note, however, that the same symbol is often used in combinatorics for a different quantity, the so-called falling factorial, for which a different symbol is used in [1].

The powerful concept of analytic continuation was developed by Weierstrass (1815–1897) and Riemann (1826–1866). We note that analytic continuation was already understood, much earlier, by Euler (1707–1783), who extensively used what we now call analytic continuation of power series. Then, there was no suitable language for Euler to transmit such ideas to his contemporaries. As a result, many of his manipulations were only understood much later [20]. For more on analytic continuation, see [3], [4], [11]–[17]—the treatments of [4] and [14] are quite thorough.

The topic of multivalued functions and their branches is generally considered to be difficult, requiring much care. An easy-to-follow, detailed discussion can be found in [17].

For the history of Hallén's and Pocklington's equations with the delta-function generator, see the Introduction of [21] and the references therein. For fine discretizations, the nonsolvability discussed in Section 3.6.1 has, of course, important consequences when one attempts to solve equations of this type numerically: see [5] and [6] for the two cases discussed in Section 3.6.1, and [21, 22], and Chapter 8 for the case of the delta-function generator.

Besides [10], standard references for MAS and related methods are [23] and [24]. The well-known Extended Integral Equation (EIE) is very similar to MAS but, at least for the simple geometry of Fig. 3.9, does not present the nonsolvability problem discussed in Section 3.6.2: see [25] and Problem 3.20. The critical distance ρ_{cri} of Eq. (3.60) is, of course, only relevant to the simple geometry of Fig. 3.9. It is, however, also pertinent to the analytic continuation of the scattered field into the region interior to the scatterer; this observation provides a bridge from the simple geometry to more complicated ones [9, 26].

3.8   PROBLEMS

3.1. Weierstrass defined the gamma function through the infinite product [1, 3]

$$\frac{1}{\Gamma(z)} = ze^{\gamma z} \prod_{n=1}^{\infty} \left[\left(1 + \frac{z}{n} \right) e^{-z/n} \right] , \tag{3.68}$$

where γ is Euler's constant.

(i) Use Eq. (3.68) together with Euler's infinite product representation of the sine function,

$$\sin z = z \prod_{n=1}^{\infty} \left(1 - \frac{z^2}{\pi^2 n^2} \right) , \tag{3.69}$$

to find a simple representation for the product $\Gamma(z)\Gamma(-z)$. Then use this expression together with the recurrence formula to show the reflection formula.

(Equation (3.69) can be found in [1]; for a proof, see [3, 11] or [13].)

3.2. **More on Euler's constant γ:**

(i) Use Eq. (3.68) and the definition of $\psi(z)$ to show $\psi(1) = -\gamma$ (Eq. (3.16)).

(ii) Show that Eq. (3.16) amounts to the following integral representation for γ,

$$\gamma = - \int_0^{\infty} e^{-t} \ln t \, dt . \tag{3.70}$$

(iii) Derive the additional integral representation

$$\gamma = - \int_0^1 \frac{e^{-t} - 1}{t} \, dt - \int_1^{\infty} \frac{e^{-t}}{t} \, dt . \tag{3.71}$$

Hint: Separate the integral in Eq. (3.70) as $\int_0^1 + \int_1^{\infty}$ and integrate by parts.

3.3. Show that $\Gamma(z)$ has no zeros and that $1/\Gamma(z)$ is an entire function.

3.4. For $n = 0, 1, 2, \ldots$, show the two relations

$$\Gamma(z - n) = \frac{(-1)^n \Gamma(z)}{(1 - z)_n},$$

$$(-n)_n = (-1)^n n!.$$

Note that the second relation can be viewed as the limiting value of of Eq. (3.17) as $z \to -n$, thus analytically continuing Eq. (3.17) to $z = 0, -1, -2, \ldots$.

3.5. Show that $\psi(z)$ has simple poles at the points $z = 0, -1, -2, \ldots$ and that the residue at any such point is -1. Then, show that

$$\lim_{z \to -n} \frac{\psi(z)}{\Gamma(z)} = (-1)^{n+1} n!, \quad n = 0, 1, 2, \ldots . \tag{3.72}$$

3.6. The well-known binomial theorem is

$$(a + b)^m = \sum_{n=0}^{m} \binom{m}{n} a^n b^{m-n}, \quad m = 0, 1, 2, \ldots ,$$

where $\binom{m}{n} = \frac{m!}{n!(m-n)!}$ are the binomial coefficients. Deduce the binomial theorem from the more general binomial expansion Eq. (3.27).

3.7. Manipulations with power series:

(i) Use Eqs. (3.22)–(3.29) to find Maclaurin series and corresponding radii of convergence for the functions $1/(\alpha z + \beta)$, $\cos^2 z$, $\ln\left[(2 + z)/(2 - z)\right]$, $\int_0^z \frac{\sin t}{t} dt$.

(ii) Expand $1/(\alpha z + \beta)$ in inverse powers of z, i.e., find its Taylor series about infinity. Where does the series converge?

(iii) If $w(z)$ is a power series in the variable z and $f(w)$ is a power series in w, then it is often possible to collect like powers of z in the composite function $f(w(z))$ in order to obtain its power series (in z). Use this principle to show that

$$\frac{1}{1 - \sin z} = 1 + z + z^2 + \frac{5}{6}z^3 + \frac{2}{3}z^4 + O(z^5), \quad \text{as} \quad z \to 0 .$$

(iv) Show that power series can be multiplied like polynomials in the sense that

$$\left(\sum_{n=0}^{\infty} \alpha_n z^n \right) \left(\sum_{n=0}^{\infty} \beta_n z^n \right) = \sum_{n=0}^{\infty} \gamma_n z^n, \quad |z| < \min(\rho_1, \rho_2) , \tag{3.73}$$

where

$$\gamma_n = \sum_{k=0}^{n} a_k b_{n-k} = \sum_{k=0}^{n} a_{n-k} b_k \,,$$

and where ρ_1, ρ_2 are the radii of convergence of the series in the left-hand side of Eq. (3.73). Show by example that the actual radius of convergence of the series in the right-hand side of Eq. (3.73) can be larger than $\min(\rho_1, \rho_2)$.

(v) Using the above rule for the multiplication of power series, show that

$$\frac{1}{\cos z} = 1 + \frac{1}{2} z^2 + \frac{5}{24} z^4 + \frac{61}{720} z^6 + O(z^8), \quad \text{as} \quad z \to 0 \,.$$

3.8.   Euler numbers E_n [1]: Define $0 = E_1 = E_3 = E_5 = \ldots$, and define E_{2n} through the power series of the secant function

$$\frac{1}{\cos z} = \sum_{n=0}^{\infty} \frac{(-1)^n E_{2n}}{(2n)!} z^{2n} \,,$$

whose first few terms were encountered in Problem 3.7.

(i) Show that

$$\frac{2e^z}{1 + e^{2z}} = \sum_{n=0}^{\infty} \frac{E_n}{n!} z^n \,.$$

(ii) Determine the radii of convergence of the above two power series.

(iii) Show that the Euler numbers can be successively calculated from

$$E_n = 1 - \sum_{k=1}^{n} \binom{n}{k} 2^{k-1} E_{n-k}, \quad n = 0, 1, 2, \ldots,$$

where, for $n = 0$, the empty sum is to be interpreted, in the usual manner, as zero.

3.9.   Hypergeometric function:

The hypergeometric function $_2F_1(a, b; c; z)$ (also denoted $F(a, b; c; z)$) is defined as the power series [1]

$$_2F_1(a, b; c; z) = \sum_{n=0}^{\infty} \frac{(a)_n (b)_n}{(c)_n} \frac{z^n}{n!} \,.$$

(i) It is usually assumed that $c \neq 0, -1, -2, \ldots$. Explain the need for this restriction.

(ii) Find the radius of convergence of the power series and compare your answer to [1].

3.10. An integral representation for the Riemann zeta function $\zeta(z)$ is [1]

$$\zeta(z) = \frac{1}{\Gamma(z)} \int\limits_0^\infty \frac{t^{z-1}}{e^t - 1}\, dt \,,$$

which is a Mellin-transform relation. Using Appendix B, determine values of z (within a vertical half-plane) for which the integral is convergent. Explain why $\zeta(z)$ is analytic for these values.

3.11. Show the equality

$$\int\limits_0^\infty t^{z-1}\cos t\, dt = (1 - z) \int\limits_0^\infty t^{z-2}\sin t\, dt \,,$$

and, with the aid of Appendix B, explain why it provides the analytic continuation of the function on the left-hand side.

3.12. Extend the considerations of Section 3.4.1 to the function z^α, where $\alpha \in \mathbb{Q}\backslash\mathbb{Z}$ (i.e., α is rational but non-integer). What happens if we encircle the origin more than once? What happens when $\alpha \in \mathbb{Z}$?

3.13. When $\alpha \in \mathbb{Z}$, verify that Eq. (3.41) reduces to a single-valued function. For a cut plane like the one in Fig. 3.5(a), show from Eq. (3.41) that z^α has a finite number of branches when $\alpha \in \mathbb{Q}\backslash\mathbb{Z}$ and an infinite number when $\alpha \in \mathbb{C}\backslash\mathbb{Q}$.

3.14. Explain the fallacies in the following sequences of equalities

$$1 = \sqrt{1} = \sqrt{(-1)(-1)} = \sqrt{-1}\sqrt{-1} = i\cdot i = -1 \,,$$
$$0 = \ln 1 = \ln\left[(-1)^2\right] = 2\ln(-1) = 2i\pi \,,$$
$$e^{i\theta} = \left(e^{i\theta}\right)^{2\pi/(2\pi)} = \left(e^{i2\pi}\right)^{\theta/(2\pi)} = (1)^{\theta/(2\pi)} = 1 \,.$$

The last equality appeared under the title "Euler's identity?" in the Spring 1989 *Newsletter of the Northeastern Section of the Mathematical Society of America* [17].

3.15. Determine conditions under which the following equalities are true

$$\sqrt{z_1 z_2} = \sqrt{z_1}\sqrt{z_2} \,,$$
$$\ln(z_1 z_2) = \ln(z_1) + \ln(z_2) \,.$$

Here, $\sqrt{}$ and $\ln$ denote principal values.

3.16. Determine the branch points of $\sqrt{1 + \sqrt{z}}$. Then, define suitable branch cuts.

3.17. Evaluate the integral

$$B = \int_0^\infty \frac{dx}{x^2 + 3x + 2}$$

by means of the auxiliary integral

$$B_2 = \int_C \frac{\ln z|_0 \, dz}{z^2 + 3z + 2},$$

where C is the closed contour of Fig. 3.11 and $\ln z|_0$ is defined in Eq. (3.38). *Hint:* First, evaluate B_2 using the residue theorem; then, show that the integrals over the large and small circles in Fig. 3.11 vanish in the limits $R \to \infty$ and $\epsilon \to 0$, where R and ϵ are the respective radii; finally, use the closed definition of the logarithm to relate the remaining two integrals (along the two horizontal paths) to the desired integral B. You can compare your solution to that of [13].

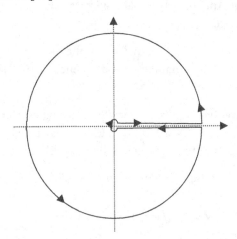

Figure 3.11: Contour C for Problems 3.17 and 3.18.

3.18. (i) If $0 < z < 1$ and $-\pi < \alpha < \pi$, use Appendix B to show that the integral

$$f(z, \alpha) = \int_0^\infty \frac{x^{z-1}}{x + e^{i\alpha}} \, dx$$

is convergent.

  (ii) By means of an auxiliary integral over the closed contour C of Fig. 3.11, show that

$$f(z, \alpha) = \frac{\pi e^{i(z-1)\alpha}}{\sin \pi z}.$$

(iii) Use an analytic continuation argument to show that the result in (ii) remains valid when $0 < \Re z < 1$.

(We note that a procedure like this can be generalized to evaluate Mellin transforms of the form

$$\int_0^\infty x^{z-1} Q(x)\, dx \, ,$$

where $Q(x)$ is a rational function with no poles on $(0, +\infty)$ and $x^z Q(x) \to 0$ both when $x \to 0$ and when $x \to +\infty$ [3].)

3.19. Hankel's loop integral [1] is

$$\frac{1}{\Gamma(z)} = \frac{1}{2\pi i} \int_C e^t t^{-z} dt \, , \tag{3.74}$$

where the contour C starts at $-\infty$ on the real axis, encircles the origin once clockwise, and returns to $-\infty$. t^{-z} attains its principal value when t crosses the positive real semi-axis, and is continuous throughout C. Use a branch of t^{-z} (in the t-plane) consistent with this definition to show that, for $\Re z > 0$, Eq. (3.74) reduces to Eq. (3.1).

3.20. For the simple geometry of Fig. 3.9, the Extended Integral Equation (EIE) is quite similar to Eq. (3.56). It is [25]

$$\int_{-\pi}^{\pi} H_0^{(1)}(kR_{cyl,aux}) J^S(\varphi_{cyl}) d\varphi_{cyl} = -\frac{I}{\rho_{cyl}} H_0^{(1)}(kR_{fil,aux}),$$

$$-\pi < \varphi_{aux} < \pi \, , \tag{3.75}$$

where $R_{cyl,aux}$ is defined in Eq. (3.57) and where

$$R_{fil,aux} = \sqrt{\rho_{fil}^2 + \rho_{aux}^2 - 2\rho_{fil}\rho_{aux} \cos \phi_{aux}} \, .$$

The left-hand side of Eq. (3.75) is the scattered field expressed in terms of the unknown $J^S(\varphi_{cyl})$, which is the true surface current density on the PEC scatterer. Equation (3.75) states that the incident field cancels the scattered field on an auxiliary surface inside the scatterer. Using the inequalities $0 < \rho_{aux} < \rho_{cri} < \rho_{cyl}$—with ρ_{cri} defined in Eq. (3.60)—show that the analytic continuation arguments of Section 3.6.2 bring up no apparent contradiction. You can compare your solution to the discussions of [25]; this reference further shows that Eq. (3.75) is solvable for any ρ_{aux} satisfying $\rho_{cri} < \rho_{aux} < \rho_{cyl}$, and determines the solution explicitly.

REFERENCES

[1] F. W. J. Olver, D. W. Lozier, R. F. Boisvert, and C. W. Clark, *Digital Library of Mathematical Functions*, National Institute of Standards and Technology from `http://dlmf.nist.gov/`, §4.2.1, §4.2.5, §4.2.7, §4.19.5, §4.22.1, §4.37(ii), Chapter 5, §15.2, §24.1-§24.5, §25.5.1. 35, 36, 46, 52, 57, 59, 60, 62, 63, 65

[2] E. C. Titchmarsh, *The Theory of Functions, 2nd Edition*. Oxford, UK: Oxford University Press, 1976, §4.21. 39

[3] E. T. Whittaker and G. N. Watson, *A Course in Modern Analysis, 4th Edition*. New York: Cambridge University Press, 1927 (reprinted, 1992), §2.3-§2.6, §5.31, §5.32, §5.501, §6.24, §12.1, §12.11. 39, 41, 43, 44, 59, 60, 65

[4] K. Knopp, *Theory of Functions, Parts I and II*. New York: Dover, 1996, §I.4, §I.22, §I.24. 41, 42, 59

[5] G. Fikioris, "The approximate integral equation for a cylindrical scatterer has no solution," *J. Electromagnetic Waves and Applications*, vol. 15, no. 9, pp. 1153–1159, September 2001. DOI: 10.1163/156939301X01075. 53, 54, 60

[6] G. Fikioris, J. Lionas, and C. G. Lioutas, "The use of the frill generator in thin-wire integral equations," *IEEE Trans. Antennas Propagat.*, vol. 51, no. 8, pp. 1847–1854, August 2003. DOI: 10.1109/TAP.2003.815412. 53, 54, 55, 60

[7] S. A. Schelkunoff, *Advanced Antenna Theory*. New York: Wiley, 1952, §5.5. 53

[8] C. A. Balanis, *Antenna Theory, Analysis and Design, 3rd Ed*. New York: Wiley, 2005, §8.3.3. 53

[9] G. Fikioris, "On two types of convergence in the Method of Auxiliary Sources," *IEEE Trans. Antennas Propagat.*, vol. 54, no. 7, pp. 2022–2033, July 2006. DOI: 10.1109/TAP.2006.877171. 55, 56, 58, 60

[10] D. I. Kaklamani and H. T. Anastassiu, "Aspects of the Method of Auxiliary Sources (MAS) in computational electromagnetics," *IEEE Antennas Propagat. Magazine.*, vol. 44, no. 3, pp. 48–64, June 2002. DOI: 10.1109/MAP.2002.1028734. 55, 60

[11] M. J. Ablowitz and A. S. Fokas, *Complex variables: Introduction and application, 2nd Ed*. Cambridge University Press, 2003. DOI: 10.1017/CBO9780511791246. 59, 60

[12] J. W. Brown and R. V. Churchill, *Complex Variables and Applications, 8th Ed*. New York: McGraw-Hill, 2004.

[13] G. F. Carrier, M. Krook, and C. E. Pearson, *Functions of a Complex Variable; Theory and Technique*. Philadelphia, PA: SIAM, 2005 DOI: 10.1137/1.9780898719116. 60, 64

[14] P. Henrici, *Applied and Computational Complex Analysis, Vol. 1*. New York: John Wiley, 1974. 59

[15] A. I. Markushevich, *Theory of Functions of a Complex Variable, 2nd Ed.* Englewood Cliffs, NJ: Chelsea Publishing Co., 1985.

[16] G. Polya and G. Latta, *Complex Variables*. New York: John Wiley, 1974.

[17] A. D. Wunsch, *Complex Variables with Applications, 3rd Ed.* Boston, MA: Pearson Education, Inc., 2005. 59, 63

[18] P. J. Davis, "Leonhard Euler's Integral: A Historical Profile of the Gamma Function," *American Mathematical Monthly*, Dec. 1959, pp. 849–869. DOI: 10.2307/2309786. 59

[19] J. Havil, *Gamma; Exploring Euler's Constant*. Princeton, NJ: Princeton Univ. Press, 2003. 59

[20] M. A. Evgrafov, "Series and Integral Representations," ch. I in M. V. Gamkrelidze (Ed.), *Analysis I: Integral Representations and Asymptotic Methods*. Berlin: Springer-Verlag, 1989. 59

[21] G. Fikioris, P. J. Papakanellos, Th. K. Mavrogordatos, N. Lafkas, and D. Koulikas, "Eliminating unphysical oscillations arising in Galerkin solutions to classical integral equations of antenna theory," *SIAM J. Appl. Math.*, vol. 71, no. 2, pp. 559–585, 2011. DOI: 10.1137/100785727. 59, 60

[22] G. Fikioris and T. T. Wu, "On the application of numerical methods to Hallen's equation," *IEEE Trans. Antennas Propagat.*, vol. 49, no. 3, pp. 383–392, March 2001. DOI: 10.1109/8.918612. 60

[23] A. Doicu, Y. Eremin, and T. Wriedt, *Acoustic and Electromagnetic Scattering using Discrete Sources*. London, UK: Academic Press, 2000. 60

[24] T. Wriedt, Ed., *Generalized Multipole Techniques for Electromagnetic and Light Scattering* (vol. 4 in Computational Methods in Mechanics). Amsterdam, The Netherlands: Elsevier, 1999. 60

[25] G. Fikioris, N. Tsitsas, and I. Psarros, "On the nature of oscillations in discretizations of the extended integral equation," *IEEE Trans. Antennas Propagat.*, vol. 59, no. 4, pp. 1415–1419, April 2011. DOI: 10.1109/TAP.2011.2109679. 60, 65

[26] G. Fikioris and I. Psarros, "On the phenomenon of oscillations in the Method of Auxiliary Sources," *IEEE Transactions Antennas Propagat*, vol. 55, no. 5, pp. 1293–1304, May 2007. DOI: 10.1109/TAP.2007.895621. 60

CHAPTER 4

Laplace's Method and Watson's Lemma

4.1 LAPLACE'S METHOD

Laplace's method gives an asymptotic approximation to certain types of integrals dependent on a large parameter: the main contribution to these integrals only comes from a small part of the integration interval, something that allows us to replace the integrand by a simpler one. The discussions that follow are heuristic and through examples, but the key ideas are widely applicable.

4.1.1 SIMPLE EXAMPLE

We will find an asymptotic approximation to

$$f(x) = \int_{-\infty}^{\infty} e^{-x\sqrt{t^2+1}}\, dt, \quad x > 0\,, \tag{4.1}$$

as $x \to +\infty$. The function $\sqrt{t^2 + 1}$ is positive throughout the integration interval, with a minimum at $t = 0$. Therefore, the integrand $g(x, t) = e^{-x\sqrt{t^2+1}}$ is hill-shaped, with a maximum at $t = 0$. Furthermore, increasing x results in a more rapid decrease away from $t = 0$ and a steeper hill shape; see Problem 4.1 for a quantification of this remark.

   The above observations are also true—and easier to depict graphically, see Fig. 4.1—for the integrand $h(x, t)$ in

$$f(x) = e^{-x} \int_{-\infty}^{\infty} e^{-x\left(\sqrt{t^2+1}-1\right)}\, dt\,. \tag{4.2}$$

In Eq. (4.2), we have re-written Eq. (4.1) so that the scaled integrand $h(x, t) = e^{-x\left(\sqrt{t^2+1}-1\right)} = e^{x}g(x, t)$ attains a maximum value—namely, the value $h(x, 0) = 1$—that is independent of x.

   As x grows, the integrand is therefore increasingly concentrated near $t = 0$. Thus, as $x \to +\infty$, the main contribution in Eq. (4.1) or Eq. (4.2) comes from smaller values of $|t|$. Therefore, the accuracy of the approximate formula

$$f(x) \sim e^{-x} \int_{-\infty}^{\infty} e^{-x\left(t^2/2\right)}\, dt \quad (x \to +\infty) \tag{4.3}$$

Figure 4.1: Plot of integrand $h(x,t) = e^{-x(\sqrt{t^2+1}-1)}$ of (4.2) as a function of t (i) for $x = 12.5$ (solid line), (ii) for $x = 50$ (dashed line), and (iii) for $x = 200$ (dot-dashed line). The equation of the horizontal line is $t = 1/e \cong 0.368$; this line is relevant to Problem 4.1.

improves as x becomes larger. Equation (4.3) results from either Eq. (4.1) or Eq. (4.2) if one replaces $\sqrt{t^2 + 1}$ by the first two terms in the small-t approximation resulting from Eq. (3.27), namely, by $1 + t^2/2$. (Using only the first term would be insufficient because the resulting integral would diverge.) The relative error in our approximation of $h(x, t)$ is small near $t = 0$, which is the region that matters. It is unimportant that this relative error is large for large values of $|t|$, because such values do not contribute—it is the absolute error that matters.

The integrand in Eq. (4.3) can be evaluated exactly by writing $\int\limits_{-\infty}^{\infty} = 2 \int\limits_{0}^{\infty}$ and then setting $xt^2/2 = u$. Thus, $\int\limits_{-\infty}^{\infty} e^{-xt^2/2} \, dt = \sqrt{2/x} \int\limits_{0}^{\infty} u^{1/2-1} e^{-u} \, du = \sqrt{2\pi/x}$, where the last integral was evaluated using the gamma function (Section 3.1). Equation (4.3) then gives

$$f(x) \sim \sqrt{\frac{2\pi}{x}} \, e^{-x} \quad (x \to +\infty) \, , \tag{4.4}$$

which is the desired asymptotic approximation. Its error (compared to numerical integration of Eq. (4.1)) is 2.9%, 0.74%, and 0.19% when $x = 12.5$, $x = 50$, and $x = 200$, respectively. Note that the scaling of Eq. (4.2) was not essential for the derivation of Eq. (4.4), an equation that can be derived directly from Eq. (4.1); scaling was helpful only for the purpose of visualizing the integrand (Fig. 4.1).

4.1.2   RELATED EXAMPLES

Following similar heuristic arguments, we can find asymptotic approximations for several related integrals.

- The integral

$$f_1(x) = \int_0^\infty e^{-x\sqrt{t^2+1}}\,dt, \quad x > 0, \tag{4.5}$$

is exactly equal to $f(x)/2$. It is thus trivial to obtain an asymptotic approximation to $f_1(x)$ directly from Eq. (4.4). It is, however, instructive for one to repeat the arguments of Section 4.1.1 independently—and to verify the answer $f_1(x) \sim \sqrt{\pi/(2x)}\,e^{-x}$ $(x \to +\infty)$—because the maximum of the normalized integrand now occurs at the *endpoint* of the integration interval. Here, we have *half* of a hill-shaped curve, the right half of Fig. 4.1.

- The maximum of the normalized integrand of

$$f_2(x, a) = \int_a^\infty e^{-x\sqrt{t^2+1}}\,dt, \quad x > 0, \quad a > 0, \tag{4.6}$$

where a is a positive constant (that is independent of x), also occurs at the left endpoint a of the integration interval. Here, we have a decreasing curve that is *part* of the right half of Fig. 4.1. The difference with $f_1(x)$ is that the derivative no longer vanishes at the endpoint. If, however, one follows the discussions of Section 4.1.1 (see also Problem 4.2), it is apparent that this difference is not essential, and that the arguments still hold: We replace the quantity $\sqrt{t^2+1}$ by the first two terms in its Taylor-series expansion about the point $t = a$, which is now the point of maximum contribution. It is easily seen that this amounts to

$$f_2(x, a) \sim e^{-x\sqrt{a^2+1}} \int_a^\infty e^{-ax(t-a)/\sqrt{a^2+1}}\,dt \quad (x \to +\infty), \tag{4.7}$$

in which, again, the integral can be evaluated exactly. The final answer is

$$f_2(x, a) \sim \frac{\sqrt{a^2+1}}{a} e^{-x\sqrt{a^2+1}} \frac{1}{x} \quad (x \to +\infty). \tag{4.8}$$

Let us stress that Eq. (4.8) holds only if a is independent of x.

- We now turn to

$$f_3(x,a) = \int_0^a e^{-x\sqrt{t^2+1}}\, dt, \quad x > 0, \quad a > 0, \tag{4.9}$$

where, once again, a is an x-independent positive constant. Since

$$f_3(x,a) = f_1(x) - f_2(x,a) = f(x)/2 - f_2(x,a), \tag{4.10}$$

and $f_2(x,a) \ll f(x)/2$ $(x \to +\infty)$, it follows that $f(x)/2$ dominates in the right-hand side of Eq. (4.10) and that the asymptotic approximations to $f(x)/2$ and $f_3(x,a)$ coincide:

$$f_3(x,a) \sim \sqrt{\frac{\pi}{2x}}\, e^{-x} \quad (x \to +\infty). \tag{4.11}$$

In other words, the asymptotic approximation of $f_3(x,a)$ is the same for any finite value of the upper limit a and remains the same when $a = +\infty$. This is so because only small values of the integration variable t matter. While it is also correct that $f_3(x,a) \sim e^{-x} \int_0^a e^{-x(t^2/2)}\, dt$ $(x \to +\infty)$, the integral in this formula does not have a *simple* closed-form evaluation when a is finite.

- If, finally, the integrand of $f_3(x,a)$ is multiplied by a simple function such as $\cos t$, one needs only to replace $\cos t$ by 1, which is the value of $\cos t$ at $t = 0$:

$$f_4(x,a) = \int_0^a \cos t\, e^{-x\sqrt{t^2+1}}\, dt \sim \sqrt{\frac{\pi}{2x}}\, e^{-x} \quad (x \to +\infty), \quad a > 0. \tag{4.12}$$

The situation is slightly different if the integrand of $f_3(x,a)$ is multiplied by $\sin t$, which vanishes at $t = 0$: in that case, one replaces $\sin t$ by t, which is the first term in the Maclaurin series of $\sin t$. When this approximation is made together with the usual approximations of Laplace's method, we once again end up with an integral that can be calculated exactly. The final result is (Problem 4.3)

$$f_5(x,a) = \int_0^a \sin t\, e^{-x\sqrt{t^2+1}}\, dt \sim \frac{e^{-x}}{x} \quad (x \to +\infty), \quad a > 0. \tag{4.13}$$

We recommend that the reader plot the normalized integrand for several large values of x (as we did in Fig. 4.1 for the case of $f(x)$). Observe that $f_5(x,a) \ll f_4(x,a)$ $(x \to +\infty)$; this was to be expected.

4.1.3   STIRLING'S FORMULA: LEADING TERM

Stirling's formula is the large-z asymptotic expansion for $\Gamma(z)$. Its first few terms were given, without proof, in Eq. (3.11). Here, for the special case of a real and positive argument, we derive the leading term using Laplace's method. We start with the definition Eq. (3.1),

$$\Gamma(x) = \int_0^\infty u^{x-1} e^{-u}\, du = \int_0^\infty e^{-[u-(x-1)\ln u]}\, du, \quad x > 0\,. \tag{4.14}$$

When $x > 1$ the integrand is hill-shaped, but the maximum occurs at the position $u = x - 1$, which is x-dependent. This difficulty is overcome by the change of variable $u = (x-1)\,y$; the resulting integrand is maximum at $y = 1$. While not necessary, it is convenient to shift the maximum to $t = 0$ via the additional change of variable $t = y - 1$. The overall change is thus $u = (x-1)(t+1)$, and this results in $\Gamma(x) = (x-1)^x e^{-(x-1)} \int_{-1}^\infty e^{-(x-1)[t-\ln(1+t)]}\, dt$. Simplification of this last expression is possible by rewriting it with $x + 1$ in place of x and then using the recurrence formula Eq. (3.5). This leads to

$$\Gamma(x) = x^x e^{-x} \int_{-1}^\infty e^{-x p(t)}\, dt\,, \tag{4.15}$$

where

$$p(t) = t - \ln(1+t)\,. \tag{4.16}$$

In Eq. (4.15), the integrand is hill-shaped with a maximum at $t = 0$. Furthermore, increasing x results in a steeper hill. Therefore, as $x \to +\infty$ the main contribution in Eq. (4.15) comes from smaller values of t and we can replace $p(t)$ by the first term in its Maclaurin expansion. By Eq. (3.29), this expansion is

$$p(t) = \sum_{n=2}^\infty \frac{(-1)^n}{n} t^n = \sum_{n=0}^\infty \frac{(-1)^n}{n+2} t^{2+n}, \quad |t| < 1\,. \tag{4.17}$$

We thus obtain the asymptotic formula

$$\Gamma(x) \sim x^x e^{-x} \int_{-1}^\infty e^{-x t^2/2}\, dt \quad (x \to +\infty)\,. \tag{4.18}$$

Because of the rapid decrease away from the maximum at $t = 0$, the lower integration limit $t = -1$ makes no difference as $x \to +\infty$ and can be replaced by $t = -\infty$:

$$\Gamma(x) \sim x^x e^{-x} \int_{-\infty}^\infty e^{-x t^2/2}\, dt \quad (x \to +\infty)\,. \tag{4.19}$$

In Section 4.1.1, we showed that the integral in Eq. (4.19) equals $\sqrt{2\pi/x}$. We thus obtain

$$\Gamma(x) \sim \sqrt{2\pi}\, x^{x-1/2} e^{-x} \quad (x \to +\infty)\,, \tag{4.20}$$

which is the desired leading term of Stirling's formula Eq. (3.11).

4.1.4   AN APPLICATION TO THE THIN-WIRE CIRCULAR-LOOP ANTENNA

We now apply the above ideas to an integral pertaining to the thin-wire, circular-loop antenna [1, 2].

Assuming an $\exp(j\omega t) = \exp(jct/k)$ time dependence, the integral equation for a thin-wire circular-loop antenna is [1]

$$
\begin{aligned}
& jk\zeta_0 b^2 \int_{-\pi}^{\pi} I(\phi') \cos(\phi - \phi')\, K(\phi - \phi')\, d\phi' \\
& + j\frac{\zeta_0}{k}\frac{\partial}{\partial\phi} \int_{-\pi}^{\pi} \frac{\partial I(\phi')}{\partial\phi'} K(\phi - \phi')\, d\phi' = r(\phi), \quad -\pi < \phi \le \pi,
\end{aligned}
\tag{4.21}
$$

where the nonsingular kernel K is

$$K(\phi) = \frac{1}{4\pi} \frac{\exp\left\{-jk\sqrt{[2b\sin(\phi/2)]^2 + a^2}\right\}}{\sqrt{[2b\sin(\phi/2)]^2 + a^2}}\,. \tag{4.22}$$

In Eq. (4.21) and Eq. (4.22), $\zeta_0 = 376.73$ Ohms, b is the loop radius, and a is the wire radius. The right-hand side $r(\phi)$ of Eq. (4.21) depends on the excitation; for the case of the delta-function generator, the frill generator, plane-wave incidence, and the finite-gap source, explicit formulas can be found in [1]. Because the wire is thin, the distance $\sqrt{[2b\sin(\phi/2)]^2 + a^2}$ in Eq. (4.22) is approximately the distance from the wire's circular axis to a point on the doughnut-shaped, perfectly conducting surface. The integral equation Eq. (4.21) can be considered as an approximation to the exact system of integral equations written in [3], which involves two perpendicular components of surface current density.

Let the Fourier-series coefficients of $I(\phi)$ be I_n so that

$$I(\phi) = \sum_{n=-\infty}^{\infty} I_n e^{jn\phi}, \quad I_n = \frac{1}{2\pi} \int_{-\pi}^{\pi} I(\phi) e^{-jn\phi}\, d\phi\,. \tag{4.23}$$

Then it is a well-known consequence of Eq. (4.21) that [1]

$$I_n = \frac{jk}{2\pi\zeta_0} \frac{r_n}{n^2 K_n - (kb)^2 (K_{n+1} + K_{n-1})/2}\,, \tag{4.24}$$

where K_n and r_n are the Fourier-series coefficients of $K(\phi)$ and $r(\phi)$. In other words, if one assumes that Eq. (4.21) has a Fourier-series solution of the form Eq. (4.23), then the coefficients are necessarily given by Eq. (4.24), *provided that the initial assumption is correct.* In this manner, one has only found a formal solution, because one has not yet addressed the question of convergence of the series in Eq. (4.23).

In Section 3.6.1 (see also [4]), we showed via an analytic continuation argument that, for the case of a *linear* antenna of length $2h$ and radius a, and for several methods of driving the antenna, Hallén's and Pocklington's equations with the approximate kernel are nonsolvable. The question we are now examining is that of the solvability of Eq. (4.21); this is a simpler question because of the explicit formal solution Eq. (4.23), and amounts to asymptotically evaluating I_n for large $|n|$.

The only nontrivial task is the asymptotic evaluation of K_n, where

$$\bar{K}_n = \frac{1}{\pi} \int_0^\pi \frac{\exp\left\{+jk\sqrt{[2b\sin(\phi/2)]^2 + a^2}\right\}}{4\pi\sqrt{[2b\sin(\phi/2)]^2 + a^2}} \cos(n\phi)\, d\phi . \tag{4.25}$$

(We initially consider the complex conjugate $\bar{K}_n$ in order to use, directly, results from [5].) Here, as a first step, we substitute the integrand from

$$\frac{\exp\left(jk\sqrt{z^2 + a^2}\right)}{4\pi\sqrt{z^2 + a^2}} = \frac{1}{2\pi^2} \int_0^\infty K_0\left(a\sqrt{\zeta^2 - k^2}\right)\cos(\zeta z)\, d\zeta , \tag{4.26}$$

where K_0 is the modified Bessel function of order zero, see Appendix A. Equation (4.26) can be deduced from Entries 2.5.25.9 and 2.5.25.15 of the first volume [6] of the monumental, five-volume table of integrals by Prudnikov, Brychkov, and Marichev. Equation (4.26) is often used in antenna theory because it can be interpreted as the Fourier inversion formula for the approximate kernel for a *straight* antenna of radius a, see Eq. (8.122) of [5] and the relevant discussions in [5]. We will use equations related to Eq. (4.26) several times in this book.

With Eq. (4.25), Eq. (4.26) gives

$$\bar{K}_n/k = \pi^{-3} \int_0^\infty dt\, K_0\left(ka\sqrt{t^2 - 1}\right) \int_0^{\pi/2} d\theta\, \cos(2kbt\sin\theta)\cos(2n\theta) , \tag{4.27}$$

where we set $\theta = \phi/2$, $t = \zeta/k$, and interchanged the order of integrations. Entry 2.5.27.1 of [6] is

$$\int_0^{\pi/2} \cos(a\sin x)\cos 2nx\, dx = \frac{\pi}{2} J_{2n}(a), \quad \mathrm{ph}\, a < \pi , \tag{4.28}$$

which gives a closed-form expression for the inner integral in Eq. (4.27). Accordingly,

$$\bar{K}_n/k = (2\pi^2)^{-1} \int_0^\infty K_0 \left(ka\sqrt{t^2-1}\right) J_{2|n|}(2kbt) \, dt \, , \tag{4.29}$$

where we replaced n by $|n|$ because the left-hand side of Eq. (4.28) is even in n. Equation (4.29) is an exact integral representation for $\bar{K}_n/k$. The integrand consists of two factors, of which only the second (i.e., $J_{2|n|}(2kbt)$) depends on n. When $|n|$ is large, the behavior of the n-dependent factor can be deduced from the asymptotic formula Eq. (A.38) or Eq. (A.40) of Appendix A: it is seen that $J_{2|n|}(2kbt)$ is a very rapidly decreasing function of $|n|$. Therefore, Laplace's method is applicable: the main contribution in Eq. (4.29) comes from large values of t, and this suggests the simple approximation $\sqrt{t^2-1} \sim t$ in the argument of the K_0. We thus obtain

$$K_n/k \sim (2\pi^2)^{-1} \int_0^\infty K_0 \left(kat\right) J_{2n}(2kbt) \, dt \, . \tag{4.30}$$

Note that the right-hand side of Eq. (4.30) is real (so that the complex conjugate is no longer necessary); this simply means that the real part of K_n/k is much larger, as $n \to +\infty$, than the imaginary part.

The reason our approximation is useful is that the integral in Eq. (4.30) can be evaluated exactly: using Entry 2.16.21.1 of the second volume [7] of the aforementioned table of integrals, we obtain

$$\frac{K_n}{k} \sim \left(4\pi^2 ka\right)^{-1} x^{2|n|} \frac{\left[\Gamma\left(|n|+\frac{1}{2}\right)\right]^2}{\Gamma(2|n|+1)} \, {}_2F_1\left(|n|+\frac{1}{2}, |n|+\frac{1}{2}; 2|n|+1; -x^2\right) \, , \tag{4.31}$$

where

$$x = 2b/a \tag{4.32}$$

with $x > 2$. In Eq. (4.31), ${}_2F_1(a, b; c; z)$ is the hypergeometric function with parameters a, b, c, see Section A.6 of Appendix A. One can also obtain Eq. (4.31) by evaluating the integral in Eq. (4.30) using the Mellin-transform method discussed in [8]. In Eq. (4.31), the parameters of the ${}_2F_1$ are linear functions, with slopes 1, 1, 2, of the large quantity $|n|$. Thus, the asymptotic approximation [9] to the ${}_2F_1$ applies, giving our final result

$$\frac{K_n}{k} \sim \frac{1}{4\pi^{3/2}ka\sqrt{|n|}} \frac{1}{(x^2+1)^{1/4}} [y(x)]^{|n|} \, , \tag{4.33}$$

where

$$y(x) = 1 - \frac{2\left(\sqrt{x^2+1}-1\right)}{x^2} \, . \tag{4.34}$$

For all $x > 2$, the function $y(x)$ is smaller than 1 and increases monotonically, with $y(\infty) = 1$. Thus, K_n decreases exponentially when $|n|$ is large.

In [1], the asymptotic formula Eq. (4.33) is checked by comparing to values of $\mathrm{Re}\,\{K_n\}$ and $\mathrm{Im}\,\{K_n\}$, obtained by integrating Eq. (4.25) numerically; such a check is especially recommended when heuristic arguments are used for the derivation of an asymptotic formula. As $|n|$ increases, the numerically obtained values of $\mathrm{Im}\,\{K_n\}$ eventually become much smaller than those for $\mathrm{Re}\,\{K_n\}$, while the relative error between Eq. (4.33) and $\mathrm{Re}\,\{K_n\}$ steadily decreases [1]. (One cannot, of course, increase $|n|$ indefinitely, as the numerical integration becomes increasingly problematic.)

Equation (4.33) can be used to determine solvability of the integral equation Eq. (4.21) as follows: Given a RHS $r(\phi)$, we determine the large-n behavior of r_n, substitute into Eq. (4.24) to find a large-n expression for I_n, and use that expression to examine the convergence of the series in Eq. (4.23). Equation (4.21) is solvable if Eq. (4.23) converges, and nonsolvable if Eq. (4.23) diverges. Because the exponentially small K_n appears in the denominator of Eq. (4.24), solvability occurs only when r_n is also exponentially small.

For the four cases mentioned above, it turns out [1] that Eq. (4.21) is nonsolvable for the cases of the delta-function generator and the finite-gap source, and solvable when the antenna is driven by a frill generator, or by an incident plane wave. For the two solvable cases, Eq. (4.33) is further used in [1] to accelerate the convergence of the Fourier-series solution in Eq. (4.23).

4.2 WATSON'S LEMMA

Some of the ideas of Laplace's method can be set in the more rigorous framework of Watson's lemma.

4.2.1 STATEMENT OF LEMMA AND MOTIVATION

Suppose that $q(t)$ is a well-behaved function that is analytic at $t = 0$. Also suppose that the Laplace transform

$$f(x) = \int_0^\infty e^{-xt} q(t)\, dt, \quad x > 0 \tag{4.35}$$

converges. Because of the factor e^{-xt}, the integrand decreases rapidly away from $t = 0$, with the decrease steeper as x increases. In accordance with Section 4.1, an asymptotic approximation to $f(x)$ as $x \to +\infty$ can be found by replacing $q(t)$ by $q(0)$; integration then yields $f(x) \sim q(0)/x$. The remainder is $\int_0^\infty e^{-xt}\left[q(t) - q(0)\right] dt$. By the same heuristic reasoning, an asymptotic approximation to this remainder is $\int_0^\infty e^{-xt} t q'(0)\, dt = q'(0)/x^2$, so that $f(x) \sim q(0)/x + q'(0)/x^2$. If continued, this process yields a formal power series for $f(x)$, which coincides with the series obtained by replacing $q(t)$ by its Maclaurin series and formally integrating the resulting expression term by term.

Watson's lemma is a theorem—to be stated here without proof—giving conditions on $q(t)$ that guarantee that the aforementioned formal series is an asymptotic power series of $f(x)$ as $x \to +\infty$. As it turns out, the said conditions are mild. It is not even essential that $q(t)$ possesses a Maclaurin series—all required is that $q(t)$ possesses a small-t asymptotic expansion consisting of powers of t, not necessarily equispaced:

Theorem 4.1 Watson's Lemma [10, 11].

Assume that the integral Eq. (4.35) converges for all sufficiently large x, with

$$q(t) \sim \sum_{k=0}^{\infty} \alpha_k t^{(k+\lambda-\mu)/\mu} \quad (t \to 0^+) , \qquad (4.36)$$

where λ and μ are positive constants. Then the series obtained by substituting Eq. (4.36) into Eq. (4.35) and integrating formally term by term yields an asymptotic expansion:

$$f(x) \sim \sum_{k=0}^{\infty} \Gamma\left(\frac{k+\lambda}{\mu}\right) \alpha_k \frac{1}{x^{(k+\lambda)/\mu}} \quad (x \to +\infty) . \qquad (4.37)$$

4.2.2 REMARKS AND EXTENSIONS

- The exponent of t in the dominant term of Eq. (4.36) (i.e., the term corresponding to $k = 0$) is greater than -1. This ensures convergence of the integral Eq. (4.35) at $t = 0$, see Rule 1 of Appendix B.

- It is noteworthy that the full asymptotic expansion of $f(x)$ is entirely determined by the small-t behavior of $q(t)$; the behavior of $q(t)$ away from $t = 0$ makes no difference.

- Since $q(t)$ need not be continuous, Eq. (4.37) also applies when the upper limit of integration in Eq. (4.35) is a finite, x-independent constant [10]. The asymptotic expansion Eq. (4.37) can then be viewed as resulting from the formal procedure of (i) substituting Eq. (4.36) into Eq. (4.35), (ii) setting the upper integration limit to infinity, and (iii) integrating term by term.

- In the form given above, Watson's lemma can also be extended to to logarithmic singularities [10, 12].

4.2.3   EXAMPLES

According to [10] and [11], Watson's lemma is probably the most frequently used method for deriving asymptotic expansions of special functions; see Problems 4.4, 4.5, and 4.6 for straightforward examples.

Since Watson's lemma is rigorous and easy to apply, it is often beneficial—when possible—to recast a quantity one wishes to expand asymptotically in the form of a Laplace transform and to then apply Watson's lemma. All integrals in Sections 4.1.1 and 4.1.2, for instance, can be handled in this manner; the two most interesting ones are $f_4(x, a)$ and $f_5(x, a)$ (Problem 4.7).

As a slightly more complicated example along these lines, let us return to the problem (Section 2.3.2) of finding a compound asymptotic approximation of $J_0(x)$ starting from the integral representation Eq. (2.9). Having already shown Eq. (2.10), it remains to find asymptotic expansions for the functions $f_1(x)$ and $f_2(x)$ defined in Eq. (2.11). Let

$$f_1(x) + i f_2(x) = \int_0^\infty \frac{e^{itx}}{\sqrt{t}\sqrt{t+2}}\, dt, \quad x > 0 . \tag{4.38}$$

By Jordan's lemma, we can shift the path of integration to the positive imaginary semi-axis (i.e, write $\int_0^\infty = \int_0^{i\infty}$) and then set $t = iy$ to obtain the Laplace transform integral

$$f_1(x) + i f_2(x) = \frac{e^{i\pi/4}}{\sqrt{2}} \int_0^\infty y^{-1/2} \left(1 + \frac{iy}{2}\right)^{-1/2} e^{-yx}\, dy \tag{4.39}$$

By Eq. (3.27), Eq. (3.17), and $\Gamma(1/2) = \sqrt{\pi}$, we have

$$\left(1 + \frac{iy}{2}\right)^{-1/2} = \frac{1}{\sqrt{\pi}} \sum_{n=0}^\infty \frac{(-1)^n \Gamma(n + 1/2)\, i^n}{n! 2^n} y^n, \quad 0 < y < 1 \tag{4.40}$$

In accordance with Watson's lemma, we now substitute Eq. (4.40) into Eq. (4.39), formally integrate term-by-term, and separate the real and imaginary parts to obtain the desired asymptotic expansions for $f_1(x)$ and $f_2(x)$. It is easy to see that the first few terms are

$$f_1(x) = \sqrt{\frac{\pi}{4x}} \left[1 + \frac{1}{8x} + O\left(\frac{1}{x^2}\right)\right], \quad f_2(x) = \sqrt{\frac{\pi}{4x}} \left[1 - \frac{1}{8x} + O\left(\frac{1}{x^2}\right)\right] \tag{4.41}$$

Thus, $J_0(x)$ has a compound asymptotic expansion consisting of the equality Eq. (2.10) and the two asymptotic expansions whose first few terms are given in Eq. (4.41). These relations can be compared to Eq. (A.30) of Appendix A.

We now show that Watson's lemma can also be used to find all terms in Stirling's formula, but the procedure is not as straightforward as one might think.

4.2.4 STIRLING'S FORMULA REVISITED AND LAGRANGE INVERSION THEOREM

In Section 4.1.3, we found the leading term in the asymptotic expansion of $\Gamma(x)$. We will now use Watson's lemma to show how to obtain as many additional terms as we desire. We have already observed that the integrand of Eq. (4.15) has a maximum at $t = 0$; the obvious next step is to write $\int_{-1}^{\infty} = \int_{-1}^{0} + \int_{0}^{\infty}$ and to change the variable to $-t$ in the first of the two integrals. This leads to

$$\Gamma(x) = x^x e^{-x} \left[\int_0^\infty e^{-xp(t)}\, dt + \int_0^1 e^{-xq(t)}\, dt \right] , \tag{4.42}$$

where

$$q(t) = -t - \ln(1 - t) , \tag{4.43}$$

and $p(t) = q(-t)$ is given in Eq. (4.16). At this point it is clear that, in order to apply Watson's lemma, we must set $u = p(t)$ in the first integral and $u = q(t)$ in the second. However, there is no simple expression for $F'(u)$ in $dt = F'(u)du$ (the meaning of our notation will soon become clear). Because Watson's lemma does not require such a simple expression—only an expansion of $F'(u)$ in powers of u—we can use the following theorem, which we give without proof.

Theorem 4.2 Extended Lagrange Inversion Theorem [10].

Suppose that

$$p(t) = p(t_0) + \sum_{n=0}^\infty p_n (t - t_0)^{\mu+n} , \tag{4.44}$$

where $\mu > 0, p_0 \neq 0$, and the series converges in a neighborhood of t_0. (For example, when μ is an integer, $p(t) - p(t_0)$ has a zero of order μ at $t = t_0$.) Let $u_0 = p(t_0)$. Then $p(t) = u$ has a solution $t = F(u)$, where

$$F(u) = t_0 + \sum_{n=1}^\infty F_n (u - u_0)^{n/\mu} \tag{4.45}$$

in a neighborhood of u_0. In Eq. (4.45), the coefficients F_n are given by

$$F_n = \frac{1}{n} \operatorname*{res}_{t=t_0} \left[\left(\frac{1}{p(t) - p(t_0)} \right)^{n/\mu} \right] , \quad n = 1, 2, \ldots . \tag{4.46}$$

For the first integral in Eq. (4.42), the substitution $u = p(t)$ together with the straightforward application of Theorem 4.2 (with the expansion of $p(t) = t - \ln(1 + t)$ given in Eq. (4.17), so that $t_0 = 0$, $\mu = 2$, and $u_0 = p(0) = 0$) yields

$$\int_0^\infty e^{-xp(t)}\, dt = \int_0^\infty e^{-xu} F'(u)\, du \,, \tag{4.47}$$

where

$$F(u) = \sum_{n=1}^\infty F_n u^{n/2} \,, \tag{4.48}$$

with

$$F_n = \frac{1}{n} \operatorname*{res}_{t=0} \left[\left(\frac{1}{t - \ln(1+t)} \right)^{n/2} \right], \quad n = 1, 2, \ldots . \tag{4.49}$$

Finding the coefficients F_n thus amounts to calculating residues. Such calculations are straight-forward but often messy, and can be greatly helped by symbolic computer programs. Explicitly, the first few coefficients given by Eq. (4.49) are found to be

$$F_1 = \sqrt{2}, \; F_2 = \frac{2}{3}, \; F_3 = \frac{1}{9\sqrt{2}}, \; F_4 = -\frac{2}{135}, \; F_5 = \frac{\sqrt{2}}{1080}, \; \ldots . \tag{4.50}$$

Equations (4.47), (4.48), and Watson's lemma then give

$$\int_0^\infty e^{-xp(t)}\, dt \sim \sum_{n=1}^\infty \frac{n}{2} F_n \int_0^\infty e^{-xu} u^{n/2-1}\, du = \sum_{n=1}^\infty F_n \, \Gamma\left(\frac{n}{2}+1\right) \frac{1}{x^{n/2}} \,, \tag{4.51}$$

where the recurrence formula Eq. (3.5) for the gamma function was used. Similarly, for the second integral in Eq. (4.42), we can show (Problem 4.10) that

$$\int_0^1 e^{-xq(t)}\, dt \sim \sum_{n=1}^\infty (-1)^{n+1} F_n \, \Gamma\left(\frac{n}{2}+1\right) \frac{1}{x^{n/2}} \,, \tag{4.52}$$

in which the F_n are those of Eq. (4.49). Equations (4.42), (4.51), and (4.52) yield the asymptotic expansion

$$\Gamma(x) = 2x^{x-1/2} e^{-x} \sum_{k=0}^\infty F_{2k+1} \, \Gamma\left(k+\frac{3}{2}\right) \frac{1}{x^k} \,, \tag{4.53}$$

which can be used together with Eqs. (4.49), (3.8), and (3.5) to obtain Stirling's formula to as many terms as one desires. The first few terms are provided in Eq. (3.11).

4.2.5   AN APPLICATION TO THE METHOD OF AUXILIARY SOURCES

The Method of Auxiliary Sources (MAS) was encountered in Section 3.6.2. In an investigation pertinent to MAS [13], we encountered the sum

$$f_n(x) = \sum_{k=1}^{n-1} \frac{x^k}{k}, \quad x \in (1, +\infty), \quad n = 2, 3, 4, \ldots . \tag{4.54}$$

The special value $f_n(1)$ can be found from Eqs. (3.15) and (3.16),

$$f_n(1) = \gamma + \psi(n) . \tag{4.55}$$

In this section, we determine the full asymptotic expansion of $f_n(x)$ as $n \to +\infty$ by finding an integral representation for $f_n(x)$ amenable to treatment by Watson's lemma. (Ref. [13] actually determines the leading term of a more general sum.) It is to be expected that $f_n(x)$ is exponentially large.

It is a consequence of Eq. (4.54) that $f_n(x)$ is an entire function with a derivative $f'_n(x)$ given by

$$f'_n(x) = \frac{x^{n-1} - 1}{x - 1} . \tag{4.56}$$

Equation (4.56) together with $f_n(0) = 0$ give

$$f_n(x) = \int_0^x \frac{t^{n-1} - 1}{t - 1} \, dt . \tag{4.57}$$

In Eq. (4.57), it is incorrect to write the integrand as $t^{n-1}/(t-1) - 1/(t-1)$ and to then integrate each term separately because both resulting integrands have poles at $t = 1$: by Rule 1 of Appendix B, a pole is a non-integrable singularity. (The original integral Eq. (4.57) converges because the integrand has a removable singularity at $t = 1$.) This difficulty can be circumvented by using the Cauchy principal value: For $a < c < b$, this is defined as [10]

$$\mathrm{PV} \int_a^b g(t) \, dt = \lim_{\varepsilon \to 0+} \left[\int_a^{c-\varepsilon} g(t) \, dt + \int_{c+\varepsilon}^b g(t) \, dt \right] , \tag{4.58}$$

provided the two integrals in the right-hand side exist for sufficiently small, positive ε (but not necessarily when $\varepsilon = 0$). This concept extends the definition of the integral: The Cauchy principal value of a convergent integral equals the integral itself, while Eq. (4.58) additionally makes sense, for example, for an integrand with a simple pole at $t = c$.

Equation (4.57) can thus be written as

$$f_n(x) = \mathrm{PV} \int_0^x \frac{t^{n-1}}{t - 1} \, dt - \mathrm{PV} \int_0^x \frac{1}{t - 1} \, dt = \mathrm{PV} \int_0^x \frac{t^{n-1}}{t - 1} \, dt - \ln(x - 1) . \tag{4.59}$$

Setting $u = \ln(x/t)$ gives

$$f_n(x) = -\ln(x-1) + x^n \, \mathrm{PV} \int_0^\infty \frac{e^{-nu}}{xe^{-u} - 1} \, du \, . \tag{4.60}$$

The integrand of Eq. (4.60) has a simple pole at $u = \ln x$. We will soon show that the logarithm in Eq. (4.60) and the contributions from large values of u can both be neglected as $n \to \infty$, so that

$$f_n(x) \sim x^n \int_0^y \frac{e^{-nu}}{xe^{-u} - 1} \, du, \qquad (n \to \infty), \quad 0 < y < \ln x \, . \tag{4.61}$$

Before applying Watson's lemma to the integral in Eq. (4.61), we provide—for clarity—an alternative derivation of Eq. (4.61): Using Eq. (4.55) and Eq. (4.56) we get

$$f_n(x) = \gamma + \psi(n) + \int_1^x \frac{t^{n-1} - 1}{t - 1} \, dt \, . \tag{4.62}$$

Integrate Eq. (4.62) by parts to obtain

$$f_n(x) = \gamma + \psi(n) + (x^{n-1} - 1)\ln(x-1) - (n-1)\int_1^x t^{n-2}\ln(t-1) \, dt \, . \tag{4.63}$$

Equation (4.63) can be rearranged to give

$$f_n(x) = \gamma + \psi(n) + (n-1)\int_1^x t^{n-2}\ln\frac{x-1}{t-1} \, dt \, , \tag{4.64}$$

which with $u = \ln(x/t)$ yields

$$f_n(x) = \gamma + \psi(n) + (n-1)x^{n-1}\int_0^{\ln x} e^{-(n-1)u}\ln\frac{x-1}{xe^{-u} - 1} \, du \, . \tag{4.65}$$

For large n, the main contribution to the integral in Eq. (4.65) comes from a narrow region near $u = 0$. We can thus replace the upper limit $\ln x$ by any smaller positive number y:

$$f_n(x) \sim \gamma + \psi(n) + (n-1)x^{n-1}\int_0^y e^{-(n-1)u}\ln\frac{x-1}{xe^{-u} - 1} \, du, \quad (n \to \infty), \quad 0 < y < \ln x \, .$$

$$\tag{4.66}$$

An integration by parts yields

$$f_n(x) \sim x^n \int_0^y \frac{e^{-nu}}{xe^{-u} - 1} \, du, \qquad (n \to \infty), \quad 0 < y < \ln x, \qquad (4.67)$$

where we neglected $\gamma + \psi(n)$ and an exponentially small boundary term, something that we will soon justify. This concludes our alternative derivation of Eq. (4.61).

To apply Watson's lemma to Eq. (4.61), we expand $(xe^{-u} - 1)^{-1}$ into a Maclaurin series. This series has the form

$$\frac{1}{xe^{-u} - 1} = \sum_{k=0}^{\infty} \alpha_k(x) \, u^k. \qquad (4.68)$$

The expansion in Eq. (4.68) is cumbersome, but it can be shown (by hand or with the aid of symbolic programs) that the first few terms are

$$\begin{aligned}
\alpha_0(x) &= \frac{1}{x - 1}, \\[2mm]
\alpha_1(x) &= \frac{x}{(x - 1)^2}, \\[2mm]
\alpha_2(x) &= \frac{x^2 + x}{2(x - 1)^3}, \\[2mm]
\alpha_3(x) &= \frac{x^3 + 4x^2 + x}{6(x - 1)^4}.
\end{aligned} \qquad (4.69)$$

Now substitute Eq. (4.68) into Eq. (4.61), set the upper limit to infinity, and formally integrate term-by-term to get the desired asymptotic expansion

$$f_n(x) \sim x^n \sum_{k=0}^{\infty} k! \alpha_k(x) \frac{1}{n^{k+1}}, \qquad (n \to \infty). \qquad (4.70)$$

Since Eq. (4.70) shows that $f_n(x)$ is exponentially large in n, our neglect of the various terms in both of our derivations of Eq. (4.61) has now been justified.

The leading term in Eq. (4.70) agrees with that given in [13]. The exponential largeness of $f_n(x)$ pertains to the near field of the nonsolvable MAS problem discussed in Section 3.6.2. This property (as well as others) implies that the near field has superdirective characteristics [13, 14].

4.3   ADDITIONAL REMARKS

Both heuristic and rigorous discussions of Laplace's method can be found in most books on asymptotic techniques, e.g., [11], [12], [15]–[18]. These references also discuss Watson's lemma (often in different forms than the one herein, see, e.g., [15]).

The basic ideas of Laplace's method and Watson's lemma can be extended, notably to the method of steepest descents and the associated saddle-point method. Given an integral to be evaluated asymptotically, the basic idea is to deform the integration path in the complex plane so as to obtain a hill-shaped integrand along the new path. These methods, which are beyond the scope of the present book, are discussed in both the mathematical and the electromagnetics literature [11], [12], [15]–[22].

Our Theorem 4.2 is called the Extended Inversion Theorem in [10]. This is a stronger version of what [10] calls the Lagrange Inversion Theorem; the latter theorem, which does not allow $p(t) - p(t_0)$ to be zero, is proved in [23].

Cauchy principal values are discussed in most books on complex variables, e.g., [17, 18].

4.4 PROBLEMS

4.1. Define the positive quantity Δt by the equation $g(x, \Delta t/2) = e^{-1}g(x, 0)$, or its equivalent $h(x, \Delta t/2) = e^{-1}h(x, 0)$, where $g(x, t)$ and $h(x, t)$ are the integrands in Eq. (4.1) and Eq. (4.2). The quantity Δt resembles the usual beamwidth in electromagnetics and measures the steepness of hill-shaped curves such as those in Fig. 4.1.

   (i) Use Fig. 4.1 to graphically estimate Δt for the three curves therein.

   (ii) Show that Δt is of the order of $1/\sqrt{x}$ for large x. More specifically, show that $\Delta t \sim \sqrt{8/x}$ as $x \to +\infty$.

4.2. Similarly to Problem 4.1, let us define t_0 ($t_0 > a$) by the equation $g(x, t_0) = e^{-1}g(x, a)$, where $g(x, t)$ is the integrand in Eq. (4.6). Show that $t_0 - a$ is of the order of $1/x$ for large x. More specifically, show that $t_0 \sim a + \frac{\sqrt{a^2+1}}{a}\frac{1}{x}$ as $x \to +\infty$. (Observe that $t_0 - a \ll \Delta t/2$ $(x \to +\infty)$.)

4.3. Verify Eq. (4.13) using Laplace's method, as outlined in Section 4.1.2.

4.4. The parabolic cylinder function $U(a, x)$ has the integral representation [10]

$$U(a, x) = \frac{e^{-x^2/4}}{\Gamma(1/2 + a)} \int_0^\infty t^{a-1/2} e^{-t^2/2 - xt}\, dt, \quad \Re a > -1/2 .$$

Determine the asymptotic expansion of $U(a, x)$ as $x \to +\infty$.

4.5. For the case $x > 0$, we briefly encountered the exponential integral $E_1(x) = \int_x^\infty e^{-t}/t\, dt$ in Section 1.3.2. Recast this integral as a Laplace transform and apply Watson's lemma to find the asymptotic expansion of $E_1(x)$ as $x \to +\infty$.

4.6.   The incomplete gamma function $\Gamma(a, x)$ is defined by the integral

$$\Gamma(a, x) = \int_x^\infty t^{a-1} e^{-t}\, dt \ ,$$

where, for simplicity, we assume $x > 0$. Show that

$$\Gamma(a, x) = x^a e^{-x} \int_0^\infty e^{-xt}(t + 1)^{a-1}\, dt \ ,$$

and then use Watson's lemma to arrive at the asymptotic expansion

$$\Gamma(a, x) \sim x^{a-1} e^{-x} \sum_{n=0}^\infty \frac{(-1)^n (1 - a)_n}{x^n} \quad (x \to +\infty) \ .$$

What happens when $a = 1, 2, \ldots$?

4.7.   Use Watson's lemma to verify Eqs. (4.12) and (4.13), and to find the next few terms.

4.8.   For $x > 0$, define

$$f(x) = \int_0^\infty \frac{\sin t}{t + x}\, dt; \quad g(x) = \int_0^\infty \frac{\cos t}{t + x}\, dt \ .$$

(i) With the aid of a contour integral, show that

$$f(x) = \int_0^\infty \frac{e^{-xt}}{t^2 + 1}\, dt; \quad g(x) = \int_0^\infty \frac{t e^{-xt}}{t^2 + 1}\, dt \ .$$

(ii) Find the asymptotic expansions of $f(x)$ and $g(x)$ as $x \to +\infty$. You can compare your answers with Eqs. (A.13) and (A.14) of Appendix A.

4.9.   Use Theorem 4.2 and Eq. (3.23) to obtain the first few terms in the Maclaurin series of arcsin z. (Newton, interestingly, encountered the converse: he had the series for arcsin z and wanted the series for sin z [23].)

4.10. Show Eq. (4.52).

REFERENCES

[1] G. Fikioris, P. J. Papakanellos, and H. T. Anastassiu, "On the use of nonsingular kernels in certain integral equations for thin-wire circular-loop antennas," *IEEE Trans. Antennas Propagat.*, vol. 56, no. 1, pp. 151–157, Jan. 2008. DOI: 10.1109/TAP.2007.913076. 74, 77

[2] G. Fikioris, P. J. Papakanellos, and H. T. Anastassiu, "Corrections to "On the use of non-singular kernels in certain integral equations for thin-wire circular-loop antennas"," *IEEE Trans. Antennas Propagat.*, vol. 58, no. 10, p. 3436, Oct. 2010.

[3] T. T. Wu, "Theory of the thin circular loop antenna," *J. Mathematical Physics*, vol. 3, no. 6, pp. 1301–1304, Nov.-Dec. 1962. DOI: 10.1063/1.1703875. 74

[4] G. Fikioris and T. T. Wu, "On the application of numerical methods to Hallen's equation," *IEEE Trans. Antennas Propagat.*, vol. 49, no. 3, pp. 383–392, March 2001. DOI: 10.1109/8.918612. 75

[5] T. T. Wu, "Introduction to linear antennas," ch. 8 in *Antenna Theory, pt. I*, R. E. Collin and F. J. Zucker, Eds. New York: McGraw-Hill, 1969. 75

[6] A. P. Prudnikov, Yu. A. Brychkov, and O. I. Marichev, *Integrals and Series: Elementary Functions, vol. 1.* Amsterdam: Gordon & Breach, 1986. 75

[7] A. P. Prudnikov, Yu. A. Brychkov, and O. I. Marichev, *Integrals and Series: Elementary Functions, vol. 2.* London: Taylor and Francis, 2002. 76

[8] G. Fikioris, *Mellin-transform method for integral evaluation: Introduction and applications to electromagnetics.* (Synthesis Lectures on Computational Electromagnetics #13). Morgan and Claypool Publishers, 2007. 76

[9] A. Erdélyi, W. Magnus, F. Oberhettinger, and F. G. Tricomi, *Higher Transcendental Functions, vol. I,* Malabar, FL: Krieger Publishing Co., 1981. (Reprint of 1953 ed.), p. 77, eqn. (16). 76

[10] F. W. J. Olver, D. W. Lozier, R. F. Boisvert, and C. W. Clark, *Digital Library of Mathematical Functions*, National Institute of Standards and Technology from http://dlmf.nist.gov/ , §1.4.24, §1.10(vii), §2.3, §8.2.2, §12.5.1. 78, 79, 80, 82, 85

[11] F. W. J. Olver, *Asymptotics and Special Functions.* Natick, MA: A. K. Peters, 1997, Chapter 3. 78, 79, 84, 85

[12] R. Wong, *Asymptotic Approximations of Integrals.* Philadelphia, SIAM, 2001, §2.2. DOI: 10.1137/1.9780898719260. 78, 84, 85

[13] P. Andrianesis and G. Fikioris, "Superdirective-type near fields in the Method of Auxiliary Sources (MAS)," *IEEE Trans. Antennas Propagat.*, vol. 60, no. 6, pp. 3056–3060, June 2012. DOI: 10.1109/TAP.2012.2194671. 82, 84

[14] S. P. Skobelev, 'Comments on "Superdirective-type near fields in the Method of Auxiliary Sources" (with authors' reply),' *IEEE Trans. Antennas Propagat.*, vol. 61, no. 4, p. 2360, April 2013. 84

[15] P. D. Miller, *Applied Asymptotic Analysis*. Providence, Rhode Island: American Mathematical Society, 2006. 84, 85

[16] C. M. Bender and S. A. Orszag, *Advanced Mathematical Methods for Scientists and Engineers; Asymptotic Methods and Perturbation Theory*. New York: Springer, 1999. DOI: 10.1007/978-1-4757-3069-2.

[17] M. J. Ablowitz and A. S. Fokas, *Complex variables: Introduction and application, 2nd Ed.* Cambridge University Press, 2003. DOI: 10.1017/CBO9780511791246. 85

[18] G. F. Carrier, M. Krook, and C. E. Pearson, *Functions of a Complex Variable; Theory and Technique*. Philadelphia, PA: SIAM, 2005. DOI: 10.1137/1.9780898719116. 84, 85

[19] C. A. Balanis, *Advanced Engineering Electromagnetics*. New York: Wiley, 1989, Appendix VI.

[20] L. B. Felsen and N. Marcuvitz, *Radiation and Scattering of Waves*. Piscataway, NJ: IEEE Press, 1994 (Reprint of 1973 Ed.) DOI: 10.1109/9780470546307.

[21] I. Andronov, D. Bouche, and Frédéric Molinet, *Asymptotic and Hybrid Methods in Electromagnetics*. London: The Institution of Electrical Engineers, 2005.

[22] D. Bouche, F. Molinet, and R. Mittra, *Asymptotic Methods in Electromagnetics*. New York: Springer, 1997, Appendix 4. DOI: 10.1007/978-3-642-60517-8. 85

[23] G. E. Andrews, R. Askey, and R. Roy, *Special Functions*. Cambridge, UK: Cambridge University Press, 1999, Appendix E. DOI: 10.1017/CBO9781107325937. 85, 86

CHAPTER 5

Integration by Parts and Asymptotics of Some Fourier Transforms

In Section 1.3.2, we found the large-x asymptotic expansion of the exponential integral $E_1(x)$ using repeated integrations by parts. In the present chapter, we present several examples illustrating that integration by parts is a quite general technique that can often bring out asymptotic endpoint contributions and explicit expressions for errors; this is true, in particular, of the large-argument behavior of some Fourier sine and cosine transforms. We then discuss certain techniques relevant to the small-argument behavior of Fourier sine and cosine transforms, and apply our methods to problems pertaining to wire antennas.

5.1 INTEGRATION BY PARTS AND LAPLACE TRANSFORMS

5.1.1 COMPLEMENTARY ERROR FUNCTION

The complementary error function erfc x is defined by the integral [1]

$$\operatorname{erfc} x = \frac{2}{\sqrt{\pi}} \int_x^\infty e^{-y^2} \, dy \, , \tag{5.1}$$

where, for simplicity, we assume $x > 0$. Since $de^{-y^2}/dy = -2ye^{-y^2}$, we can integrate by parts to obtain

$$\operatorname{erfc} x = -\frac{1}{\sqrt{\pi}} \int_x^\infty \frac{1}{y} \frac{de^{-y^2}}{dy} \, dy = \frac{1}{\sqrt{\pi}} \frac{e^{-x^2}}{x} - \frac{1}{\sqrt{\pi}} \int_x^\infty \frac{e^{-y^2}}{y^2} \, dy \, . \tag{5.2}$$

The integral in the right-hand side of Eq. (5.2) satisfies

$$\frac{1}{\sqrt{\pi}} \int_x^\infty \frac{e^{-y^2}}{y^2} \, dy < \frac{1}{\sqrt{\pi} \, x^2} \int_x^\infty e^{-y^2} \, dy = \frac{\operatorname{erfc} x}{2 \, x^2} \ll \operatorname{erfc} x \quad (x \to +\infty) \, , \tag{5.3}$$

so that the first (boundary) term in the right-hand side of Eq. (5.2) is an asymptotic approximation to erfc x,

$$\text{erfc } x \sim \frac{1}{\sqrt{\pi}} \frac{e^{-x^2}}{x} \quad (x \to +\infty) . \tag{5.4}$$

This process can be repeated to obtain the full asymptotic expansion of erfc x. To obtain results with wider applicability, we carry this out after first writing Eq. (5.1) as a Laplace transform by setting $t = y^2/x^2 - 1$. We obtain

$$\text{erfc } x = \frac{xe^{-x^2}}{\sqrt{\pi}} I\left(x^2\right) , \tag{5.5}$$

where the Laplace transform $I(x)$ is

$$I(x) = \int_0^\infty q(t)e^{-xt} \, dt, \quad x > 0 \tag{5.6}$$

with

$$q(t) = \frac{1}{\sqrt{1+t}} . \tag{5.7}$$

Integrating Eq. (5.6) by parts N times we obtain

$$I(x) = \sum_{n=0}^{N-1} \frac{q^{(n)}(0)}{x^{n+1}} + r_N(x), \quad N = 1, 2, \dots , \tag{5.8}$$

where

$$r_N(x) = \frac{1}{x^N} \int_0^\infty e^{-xt} q^{(N)}(t) \, dt . \tag{5.9}$$

From Eq. (5.7), it is apparent that $\left|q^{(N)}(t)\right|$ is a decreasing function of t for all N. It follows that

$$|r_N(x)| \leq \frac{\left|q^{(N)}(0)\right|}{x^N} \int_0^\infty e^{-xt} \, dt = \frac{\left|q^{(N)}(0)\right|}{x^{N+1}} . \tag{5.10}$$

Therefore, for all N, the $r_N(x)$ in Eq. (5.8) is of the order of the first neglected term. Thus,

$$I(x) \sim \sum_{n=0}^\infty \frac{q^{(n)}(0)}{x^{n+1}} \quad (x \to \infty) . \tag{5.11}$$

Setting $s = -1/2$ in Eq. (3.27) and comparing with Eq. (5.7), it is seen that $q^{(n)}(0) = (-1)^n (1/2)_n$, where we use the usual notation for Pochhammer's symbol, defined in Eq. (3.17).

Equations (5.5) and (5.11) thus give our final asymptotic expansion

$$\operatorname{erfc} x \sim \frac{1}{\sqrt{\pi}} \frac{e^{-x^2}}{x} \sum_{n=0}^{\infty} (-1)^n \left(\frac{1}{2}\right)_n \frac{1}{x^{2n}} \quad (x \to \infty).$$ (5.12)

The terms in Eq. (5.12) are the contributions from the endpoint $t = 0$ to the integral in Eq. (5.6), or from the endpoint $y = x$ to the integral in Eq. (5.1). The result in Eq. (5.12) coincides with what we obtain by applying Watson's lemma to Eq. (5.6): to see this, note that $\operatorname{erfc} x = \pi^{-1/2}\Gamma\left(1/2, x^2\right)$ and compare with the result in Problem 4.6.

Equation (5.10) reveals an additional desirable property of Eq. (5.12): the magnitude of the error $r_N(x)$ is *numerically* (not asymptotically) smaller than the magnitude of the first neglected term $q^{(N)}(0)/x^{N+1}$. This property—which is often true of integrals of the form Eq. (5.6) [2]—can help us truncate the divergent series Eq. (5.12) in an optimal manner simply by looking at the terms: we stop when the *next* term is smallest. See [3] for further discussions on the optimal truncation of asymptotic expansions.

A further desirable property of Eq. (5.12) is that successive partial sums alternate about the true value. That is, if a partial sum overestimates $\operatorname{erfc} x$, then the next partial sum underestimates $\operatorname{erfc} x$. This is true because $q^{(0)}(t)$, $q^{(1)}(t)$, $q^{(2)}(t)$,... alternate in sign, so that the $r_N(x)$ also alternate in sign by Eq. (5.9). Again, this is often true of integrals of the form Eq. (5.6) [2].

5.1.2 REMARKS

- To obtain Eq. (5.9), we invoked the property that all $\left|q^{(N)}(t)\right|$ are decreasing functions of t. For Laplace transforms, the method of integration by parts "works" under less restrictive conditions. By this, we mean that integrals of the form Eq. (5.6) can possess the asymptotic expansion Eq. (5.11) under milder conditions on $q(t)$ and its derivatives. Such conditions can be found in the literature [1, 2, 4]; the last two references also give general discussions on the integration-by-parts method for the case of integrals that are not Laplace transforms.

- One integration by parts produces two boundary terms and a new integral. It is obvious—but worth stressing—that all three of these terms must be finite. Consider, for example, the problem of finding a large-x approximation of the error function $\operatorname{erf} x$, which is defined by the integral [1]

$$\operatorname{erf} x = \frac{2}{\sqrt{\pi}} \int_0^x e^{-y^2} \, dy, \quad x > 0.$$ (5.13)

Direct integration by parts similar to that in Eq. (5.2) here does not work because neither the boundary term at $y= 0$ nor the new integral exist. The difficulty can be circumvented by writing $\int_0^x = \int_0^\infty - \int_x^\infty$, leading (Problem 5.1) to

$$\operatorname{erf} x = 1 - \operatorname{erfc} x,$$ (5.14)

and then using the asymptotic expansion Eq. (5.12) already found for erfc x. Note that all terms in Eq. (5.12) are exponentially smaller than the term 1 in Eq. (5.14); see the review article [5] for far-reaching discussions on exponentially small terms.

- Suppose that both boundary terms produced by an integration by parts are finite and nonzero. We obtain a valid asymptotic approximation when one of the two boundary terms is much larger, as $x \to \infty$, than both the new integral and the other boundary term. For example, integration by parts easily leads to the asymptotic approximation

$$\frac{2}{\sqrt{\pi}} \int_{\sqrt{x}}^{x} e^{-y^2} \, dy \sim \frac{1}{\sqrt{\pi}} \frac{e^{-x}}{\sqrt{x}} \quad (x \to +\infty) \, , \tag{5.15}$$

which is the contribution from the lower limit $y = \sqrt{x}$. Additional discussions along these lines can be found in [3].

5.2   INTEGRATION BY PARTS AND FOURIER TRANSFORMS

The main focus of Section 5.1 was Laplace transforms. In this section, we consider examples involving Fourier sine and cosine transforms.

5.2.1   SIMPLE EXAMPLE: RIEMANN-LEBESGUE LEMMA

Consider the integral

$$F_1(x, \alpha) = \int_0^{\alpha} \sqrt{t} \cos(xt) \, dt, \quad x > 0, \quad \alpha > 0 \, , \tag{5.16}$$

where α is constant independent of x. We wish to determine the first few terms in the large-x asymptotic expansion of $F_1(x, \alpha)$. An integration by parts brings out the dominant contribution from the endpoint $x = a$,

$$F_1(x, \alpha) = \frac{1}{x} \int_0^{\alpha} \sqrt{t} \frac{d \sin(xt)}{dt} \, dt = \frac{\sqrt{\alpha} \sin(x\alpha)}{x} - \frac{1}{2x} \int_0^{\alpha} \frac{\sin(xt)}{\sqrt{t}} \, dt \, . \tag{5.17}$$

As $x \to \infty$, the new integral in Eq. (5.17) can be asymptotically estimated with the aid of the Riemann-Lebesgue lemma, which we give without proof:

Theorem 5.1   Riemann-Lebesgue Lemma [1, 2].
For $f(t)$ piecewise continuous on $[a, b]$ and real x,

$$\int_a^b f(t)e^{ixt}\, dt \to 0, \quad \text{as } x \to \infty . \tag{5.18}$$

Equation (5.18) continues to apply if either a or b or both are infinite and/or $f(t)$ has finitely many singularities in (a, b), provided that the integral converges uniformly at a, b, and the singularities for all sufficiently large x.

The Riemann-Lebesgue lemma shows that the integral in Eq. (5.17) is smaller, as $x \to \infty$, than the boundary term. We have thus arrived at the asymptotic approximation

$$F_1(x, \alpha) = \frac{\sqrt{\alpha} \sin{(x\alpha)}}{x} + o\left(\frac{1}{x}\right) \quad (x \to \infty) . \tag{5.19}$$

For generalizations of the procedure that led to (5.19) to Fourier integrals of the form $\int_a^b e^{ixt} q(t)\, dt$ and $\int_a^\infty e^{ixt} q(t)\, dt$, see [1, 2, 4].

5.2.2   REMARKS ON THE LEMMA

- Our Theorem 5.1 is the Riemann-Lebesgue lemma as found in [1, 2]. Other statements of the Lemma [6, 7] assume absolute integrability of $f(t)$ (rather than uniform convergence); see [2] for comparative discussions.

- While the proof of the Riemann-Lebesgue lemma is rather involved, it is easy to understand the lemma intuitively: As x grows, the real and imaginary parts of the integrand become increasingly oscillatory so that the variations of $f(t)$ over a period are unimportant. Thus, all positive portions of the integrand will tend to cancel with their adjacent negative portions.

- In the case of infinite integration intervals, it is very important to check the uniform convergence of the integral (or the absolute integrability of $f(t)$) before applying the lemma. As an example, the lemma does not apply to $F_2(x) = \int_0^\infty \sin{(xt)}/t\, dt$ because the integral is not uniformly convergent. Indeed, $F_2(x) = \text{Si}(\infty) = \pi/2$ for all positive x (set $y = xt$ and use Eq. (A.11) of Appendix A), so that $F_2(x)$ is nonzero as $x \to \infty$.

5.2.3   SIMPLE EXAMPLE CONTINUED

We return to the simple example $F_1(x, \alpha)$ of Section 5.2.1 to find the next few terms. Direct integration by parts of the integral $\int_0^\alpha \sin{(xt)}/\sqrt{t}\, dt$ in Eq. (5.17) fails because the boundary term at $t = 0$ does not exist. To avoid the difficulty, we proceed as we did with Eq. (5.13): We write

$$F_1(x, \alpha) = \frac{\sqrt{\alpha} \sin (x\alpha)}{x} - \frac{1}{2x} \int\limits_0^\infty \frac{\sin (xt)}{\sqrt{t}} \, dt + \frac{1}{2x} \int\limits_\alpha^\infty \frac{\sin (xt)}{\sqrt{t}} \, dt \,. \qquad (5.20)$$

Since the lower limit of the last integral is nonzero, it can be integrated by parts:

$$\frac{1}{2x} \int\limits_\alpha^\infty \frac{\sin (xt)}{\sqrt{t}} \, dt = \frac{1}{2x^2} \frac{\cos x\alpha}{\sqrt{\alpha}} - \frac{1}{4x^2} \int\limits_\alpha^\infty \frac{\cos (xt)}{t^{3/2}} \, dt = \frac{1}{2x^2} \frac{\cos x\alpha}{\sqrt{\alpha}} + o\left(\frac{1}{x^2}\right) \,, \qquad (5.21)$$

where the last (asymptotic) relation follows from the Riemann-Lebesgue lemma. To evaluate the other integral in Eq. (5.20), we use

$$\int\limits_0^\infty y^{z-1} \sin y \, dy = \sin \left(\frac{\pi z}{2}\right) \Gamma(z), \quad -1 < \Re z < 1 \,. \qquad (5.22)$$

For later use, we also note the similar integral

$$\int\limits_0^\infty y^{z-1} \cos y \, dy = \cos \left(\frac{\pi z}{2}\right) \Gamma(z), \quad 0 < \Re z < 1 \,. \qquad (5.23)$$

Equations (5.22) and (5.23), which give the Mellin transforms of $\sin y$ and $\cos y$, can be found in [1]. The reader is asked to show the two equations in Problem 5.4.

Let us resume our discussion of $F_1(x, \alpha)$. By Eq. (5.22), we have

$$\frac{1}{2x} \int\limits_0^\infty \frac{\sin (xt)}{\sqrt{t}} \, dt = \frac{1}{2x^{3/2}} \int\limits_0^\infty \frac{\sin y}{\sqrt{y}} \, dy = \frac{\sqrt{\pi}}{2\sqrt{2} x^{3/2}} \,. \qquad (5.24)$$

Combining Eqs. (5.20), (5.21), and (5.24) we obtain the three-term asymptotic approximation

$$F_1(x, \alpha) = \frac{\sqrt{\alpha} \sin (x\alpha)}{x} - \frac{\sqrt{\pi}}{2\sqrt{2} x^{3/2}} + \frac{1}{2x^2} \frac{\cos (x\alpha)}{\sqrt{\alpha}} + o\left(\frac{1}{x^2}\right) \quad (x \to \infty) \,. \qquad (5.25)$$

5.2.4 EXAMPLE WITH ZERO BOUNDARY TERMS

Let us apply integration by parts to

$$F_2(x, \alpha) = \int\limits_0^\infty \frac{\cos (xt)}{t^2 + \alpha^2} \, dt, \quad x > 0, \quad \alpha > 0 \,. \qquad (5.26)$$

It is simpler to first write

$$F_2(x, \alpha) = \frac{1}{2} \int_{-\infty}^{\infty} \frac{\cos(xt)}{t^2 + \alpha^2} \, dt \, . \tag{5.27}$$

Integrating by parts once gives boundary terms that are both zero. We obtain

$$F_2(x, \alpha) = \frac{1}{x} \int_{-\infty}^{\infty} \frac{t \sin(xt)}{(t^2 + \alpha^2)^2} \, dt \, , \tag{5.28}$$

where we used $d(t^2 + \alpha^2)^{-1}/dt = -2t/(t^2 + \alpha^2)^{-2}$. The integral in Eq. (5.28) is convergent by Appendix B and $o(1)$ as $x \to \infty$ by the Riemann-Lebesgue lemma. Therefore, Eq. (5.28) shows that $F_2(x, \alpha) = o(1/x)$ as $x \to \infty$. A second integration by parts gives

$$F_2(x, \alpha) = \frac{1}{x^2} \int_{-\infty}^{\infty} \cos(xt) \left[\frac{1}{(t^2 + \alpha^2)^2} - \frac{4t^2}{(t^2 + \alpha^2)^3} \right] dt \, , \tag{5.29}$$

implying, similarly, that $F_2(x, \alpha) = o(1/x^2)$. Readers should convince themselves that this process can be continued indefinitely so that

$$F_2(x, \alpha) = o\left(\frac{1}{x^n} \right), \quad (x \to +\infty) \quad n = 0, 1, 2, \dots \, . \tag{5.30}$$

Since n in Eq. (5.30) is arbitrary, we have reached the conclusion that all terms in the asymptotic power series of $F_2(x, \alpha)$ are zero. This means that $F_2(x, \alpha)$ decays faster than any power of $1/x$ when x is large. While this conclusion does not aid in obtaining numerical values for $F_2(x, \alpha)$, it tells us that evaluation of $F_2(x, \alpha)$ by numerically integrating Eq. (5.26) is difficult: the value of the integral is much smaller than the values of the oscillating integrand—thus, the smallness is due to cancellation, something that is always a sign of trouble. (A conclusion of this type might be of further importance in applications; we will give an example in Section 5.4.1 and Chapter 8.)

In this simple example, our conclusion is readily verified because Eq. (5.26) can be calculated exactly (e.g., by contour integration, Problem 5.7). It turns out that $F_2(x, \alpha)$ is exponentially small,

$$F_2(x, \alpha) = \frac{\pi}{2\alpha} e^{-\alpha x} \, ; \tag{5.31}$$

thus, $F_2(x, \alpha)$ indeed decays faster than any power of $1/x$. A similar behavior is exhibited by the Fourier cosine transform

$$F_3(x, \alpha) = \int_0^{\infty} \frac{\cos(xt)}{\sqrt{t^2 + \alpha^2}} \, dt, \quad x > 0, \quad \alpha > 0 \, , \tag{5.32}$$

see Problem 5.10 (this problem makes use of several topics dealt with in this book and is particularly instructive).

5.3   MORE ON FOURIER TRANSFORMS

In this section, we turn to the behavior of the Fourier cosine transform $F_3(x, \alpha)$ of Eq. (5.32) for the case where x is *small*. It is natural to first try setting $x = 0$, but the resulting integral diverges at $t = \infty$, see Rule 3 of Appendix B. To circumvent this, we try to add and subtract $1/t$, which is the large-t behavior of the function $1/\sqrt{t^2 + a^2}$ multiplying the cosine. However, $1/t$ is nonintegrable at the lower limit $t = 0$, so we split then integral before doing so:

$$F_3(x, \alpha) = \int\limits_0^A \frac{\cos(xt)}{\sqrt{t^2 + \alpha^2}}\, dt + \int\limits_A^\infty \cos(xt) \left[\frac{1}{\sqrt{t^2 + \alpha^2}} - \frac{1}{t}\right] dt + \int\limits_A^\infty \frac{\cos(xt)}{t}\, dt\,, \qquad (5.33)$$

where A is any positive number that is independent of x ($A = 1$ will do). In Eq. (5.33), the first integral is $O(1)$ as $x \to 0$. By Eq. (3.27),

$$\frac{1}{\sqrt{t^2 + \alpha^2}} - \frac{1}{t} = \frac{t - t\sqrt{1 + (\alpha/t)^2}}{t\sqrt{t^2 + \alpha^2}} = \frac{O(1/t)}{O(t^2)} = O\left(\frac{1}{t^3}\right) \quad (t \to \infty)\,. \qquad (5.34)$$

Therefore, the second integral is also $O(1)$ as $x \to 0$. The third integral in Eq. (5.33) can be calculated explicitly by setting $y = xt$:

$$\int\limits_A^\infty \frac{\cos(xt)}{t}\, dt = \int\limits_{Ax}^\infty \frac{\cos y}{y}\, dy = -\mathrm{Ci}(Ax) = \ln(1/x) + O(1) \quad (x \to 0^+)\,, \qquad (5.35)$$

where we used the definition Eq. (A.2) of the cosine integral Ci and the small-argument formula for Ci, Eq. (A.8). We have thus found an asymptotic approximation to $F_3(x, \alpha)$, as well as the order of the remainder,

$$F_3(x, \alpha) = \ln\frac{1}{x} + O(1), \quad (x \to 0^+)\,. \qquad (5.36)$$

Our method of deriving Eq. (5.36) shows that the leading small-x behavior of the Fourier cosine transform $F_3(x, \alpha)$ is completely determined from the large-t behavior of the defining integral Eq. (5.32).

It is somewhat simpler to treat the Fourier sine transform of the same function,

$$F_4(x, \alpha) = \int\limits_0^\infty \frac{\sin(xt)}{\sqrt{t^2 + \alpha^2}}\, dt, \quad x > 0, \quad \alpha > 0\,, \qquad (5.37)$$

which can be expected to be much smaller than $F_3(x, \alpha)$ as $x \to 0^+$. Here, there is no need to introduce a constant A:

$$F_4(x,\alpha) = \int_0^\infty \frac{\sin(xt)}{t}\, dt + \int_0^\infty \sin(xt)\left[\frac{1}{\sqrt{t^2+\alpha^2}} - \frac{1}{t}\right] dt = \int_0^\infty \frac{\sin y}{y}\, dy + o(1)$$
$$= \frac{\pi}{2} + o(1) \quad (x \to 0^+).$$

(5.38)

In (5.38), the integral $\int_0^\infty (\sin y / y)\, dy$ was found as $\mathrm{Si}\,(+\infty) = \pi/2$, see Eq. (A.11), or by taking the limit in Eq. (5.22).

The above two integrals are Fourier cosine and sine transforms of the form

$$\bar{f}_C(x) = \int_0^\infty f(t)\cos(xt)\, dt, \quad \bar{f}_S(x) = \int_0^\infty f(t)\sin(xt)\, dt,$$

(5.39)

where

$$f(t) \sim \frac{1}{t^\beta} \quad \text{as} \quad t \to +\infty \quad (0 < \beta \le 1).$$

(5.40)

The key ideas of the procedure we followed are to add and subtract the large-t behavior of $f(t)$; to then exactly evaluate the integral that is independent of $f(t)$ (using the definition of Si or Ci in Appendix A in the case $\beta = 1$, and Eq. (5.22) or Eq. (5.23) in the case $0 < \beta < 1$). In the case $\beta = 1$ only, we additionally use the leading, large-x behavior of Si or Ci from Eqs. (A.11)–(A.14). This method often works, i.e, it often gives the leading, small-x behavior of Eq. (5.39). When carrying it out, it is important to check that all arising integrals converge—if not, a simple modification such as the introduction of the constant A in the case of $F_3(x, \alpha)$ may do the trick—and to keep track of the order, as $x \to 0^+$, of all discarded terms. A simple Fourier cosine transform that can be treated in this manner is in Problem 5.11. More complicated integrals to which our procedure is relevant are in Problems 5.12 and 5.13.

5.4   APPLICATIONS TO WIRE ANTENNAS

In this section, we give some applications of the previously developed methods to wire antennas. We first use our methods to point out important differences between the Fourier transforms of the two kernels (exact and approximate) in Hallén's and Pocklington's equations. In this section, a and k respectively stand for the antenna radius a and wavenumber k, and our time dependence is $e^{-i\omega t}$ where $k = \omega/c$.

5.4.1 ON THE KERNELS OF HALLÉN'S AND POCKLINGTON'S EQUATIONS

We encountered Hallén's and Pocklington's equations with the approximate kernel $K(z)$ in Section 3.6.1. By Eq. (3.51), the Fourier transform $\bar{K}(\zeta)$ of this kernel is

$$\bar{K}(\zeta) = \int\limits_{-\infty}^{\infty} K(z)e^{i\zeta z}\,dz = \int\limits_{-\infty}^{\infty} \frac{1}{4\pi} \frac{\exp\left(ik\sqrt{z^2 + a^2}\right)}{\sqrt{z^2 + a^2}} e^{i\zeta z}\,dz, \quad \zeta \in \mathbb{R}. \tag{5.41}$$

By Rule 5 of Appendix B, the integral in Eq. (5.41) is convergent as long as $\zeta \neq \pm k$. It is easy to see that all derivatives of $K(z)$ are linear combinations of terms of the form

$$\frac{\exp\left(ik\sqrt{z^2 + a^2}\right)}{\left(\sqrt{z^2 + a^2}\right)^n} \frac{z}{\sqrt{z^2 + a^2}}, \quad n = 1, 2, \dots, \tag{5.42}$$

which are $O\left(e^{ik|z|}/|z|\right)$ as $z \to \pm\infty$. Therefore, successive integrations by parts will produce new integrals that are convergent and boundary terms that are all equal to zero. As in Section 5.2.3, we conclude that

$$\bar{K}(\zeta) = o\left(\frac{1}{\zeta^n}\right), \quad (\zeta \to \pm\infty) \quad n = 0, 1, 2, \dots, \tag{5.43}$$

so that, for large ζ, $\bar{K}(\zeta)$ decays faster than any power of ζ. It is easy to verify this result by explicit calculation of the integral Eq. (5.41). Using Entries 2.5.25.9 and 2.5.25.15 of [8], we obtain

$$\overline{K}(\zeta) = \begin{cases} \frac{i}{4}H_0^{(1)}\left(a\sqrt{k^2 - \zeta^2}\right), & |\zeta| < k, \\ \frac{1}{2\pi}K_0\left(a\sqrt{\zeta^2 - k^2}\right), & |\zeta| > k, \end{cases} \tag{5.44}$$

where $H_0^{(1)}$ and K_0 are the Hankel and modified Bessel functions, discussed in Appendix A. Equation A.49 implies that $\bar{K}(\zeta)$ is exponentially small, thus verifying Eq. (5.43).

Let us additionally note that (i) Eq. (4.26) is the inverse Fourier relation of Eq. (5.44); (ii) the two expressions in Eq. (5.44) are analytic continuations of one another—this is discussed in detail in [9]; (iii) by Eq. (A.47), the singularities of $\bar{K}(\zeta)$ at $\zeta = \pm k$ are logarithmic and thus integrable.

We now find the large-ζ behavior for the *exact* kernel $K_{\text{ex}}(z)$. This kernel is given by [9, 10]

$$K_{ex}(z) = \frac{1}{8\pi^2} \int\limits_{-\pi}^{\pi} \frac{\exp\left(ik\sqrt{z^2 + 4a^2\sin^2\frac{\theta}{2}}\right)}{\sqrt{z^2 + 4a^2\sin^2\frac{\theta}{2}}}\,d\theta. \tag{5.45}$$

The Fourier transform $\bar{K}_{ex}(\zeta)$ can be calculated by substituting Eq. (5.45) into

$$\bar{K}_{ex}(\zeta) = \int_{-\infty}^{\infty} K_{ex}(z) e^{i\zeta z}\, dz \,, \tag{5.46}$$

and interchanging the orders of integration. The resulting inner integral is like that in Eq. (5.41) with $2a \left|\sin\left(\theta/2\right)\right|$ in place of a and can thus be evaluated with the aid of Eq. (5.44). We then evaluate the resulting integral using Entries 2.12.4.1, 2.13.3.1, and 2.16.3.1 of [11]. The result is

$$\bar{K}_{ex}(\zeta) = \begin{cases} \dfrac{i}{4} J_0\left(a\sqrt{k^2 - \zeta^2}\right) H_0^{(1)}\left(a\sqrt{k^2 - \zeta^2}\right), & |\zeta| < k \,, \\[2ex] \dfrac{1}{2\pi} I_0\left(a\sqrt{\zeta^2 - k^2}\right) K_0\left(a\sqrt{\zeta^2 - k^2}\right), & |\zeta| > k \,. \end{cases} \tag{5.47}$$

As with Eq. (5.44), the two expressions in Eq. (5.47) are analytic continuations of one another and the singularities of $\bar{K}_{ex}(\zeta)$ at $\zeta = \pm k$ are logarithmic and integrable. Equations (5.47), (A.48), and A.49 (or, more directly, the large-argument asymptotic expansion of the product $I_0 K_0$ [1]) now show that

$$\bar{K}_{ex}(\zeta) = \frac{1}{4\pi a}\frac{1}{|\zeta|} + O\left(\frac{1}{|\zeta|^3}\right), \quad (\zeta \to \pm\infty) \,. \tag{5.48}$$

Thus, $\bar{K}_{ex}(\zeta)$ decays only as $1/\left|\zeta\right|$ and behaves in stark contrast to $\bar{K}(\zeta)$. The smallness of $\bar{K}(\zeta)$, Eq. (5.43), is reminiscent of the smallness of the quantity K_n relevant to the circular-loop antenna, see Eq. (4.33). In Section 4.1.4, we saw that the smallness of K_n implies the nonsolvability of the integral equation for the circular-loop antenna driven by a delta-function generator. In Chapter 8, we will show that the smallness of $\bar{K}(\zeta)$ similarly implies the nonsolvability of the integral equation for the antenna of *infinite length*. (For the case of finite length, nonsolvability was shown in Section 3.6.1, but the arguments there are not valid when $h = \infty$.)

It is well-known that the exact kernel $K_{ex}(z)$ has a logarithmic singularity at $z = 0$. Interestingly, this can be viewed as a consequence of Eq. (5.48), the Fourier inversion formula and the methods of Section 5.3: The inverse relation of Eq. (5.46) is

$$K_{ex}(z) = \frac{1}{2\pi}\int_{-\infty}^{\infty} \bar{K}_{ex}(\zeta) e^{-i\zeta z}\, dz = \frac{1}{\pi}\int_{0}^{\infty} \bar{K}_{ex}(\zeta) \cos\left(\zeta z\right)\, dz \,, \tag{5.49}$$

so that $K_{ex}(z)$ is the Fourier cosine transform of a function behaving like $1/\left(4\pi^2 a\zeta\right)$ for large, positive ζ. Adding and subtracting this behavior and working as in Section 5.3, we can show that

$$K_{ex}(z) \sim \frac{1}{4\pi^2 a}\ln\frac{1}{z} \quad (z \to 0^+) \,. \tag{5.50}$$

A more direct way to show Eq. (5.50) is through Eq. (5.45) (Problem 5.14). We now explore a more interesting consequence of Eq. (5.48).

5.4.2   BEHAVIOR OF CURRENT NEAR DELTA-FUNCTION GENERATOR

For a finite antenna of length $2h$ center-driven by a delta-function generator at $z = 0$, and for the case of the exact kernel, Hallén's equation is given in Eq. (3.50) with $K(z)$ replaced by $K_{\mathrm{ex}}(z)$. Apart from a multiplicative constant, both sides of that equation represent the vector potential, compare to Eq. (1.1). The equation is easily specialized to the case $h = \infty$ of an antenna infinite in length: As $z \to \pm\infty$, the right-hand side must represent an outgoing wave. This leads to $C = V/(2\zeta_0)$ and the desired equation thus is

$$\int_{-\infty}^{\infty} K_{ex}(z - z')I(z')\ dz' = \frac{V}{2\zeta_0}e^{ik|z|}, \quad z \in \mathbb{R}, \tag{5.51}$$

where $I(z)$ is the unknown current on the infinite antenna. Integral equations that have the form of Eq. (5.51) can be solved by exploiting the convolution property of the Fourier transform: For $\Im k > 0$ (so that the surrounding medium is slightly lossy), Eq. (5.51) implies

$$\overline{I}(\zeta)\,\overline{K}_{ex}(\zeta) = \frac{ikV}{\zeta_0}\frac{1}{k^2 - \zeta^2}, \tag{5.52}$$

where $\overline{I}(\zeta)$ is the Fourier transform of $I(z)$. Taking the inverse Fourier transform, we obtain

$$I(z) = \frac{ikV}{2\pi\zeta_0}\int_{-\infty}^{\infty}\frac{1}{(k^2 - \zeta^2)\,\overline{K}_{ex}(\zeta)}e^{-i\zeta z}\ d\zeta = \frac{ikV}{\pi\zeta_0}\int_{0}^{\infty}\frac{\cos(\zeta z)}{(k^2 - \zeta^2)\,\overline{K}_{ex}(\zeta)}\ d\zeta. \tag{5.53}$$

Equation (5.53), which originally holds for $\Im k > 0$, can be analytically continued to real and positive k if the path of integration passes below the point $\zeta = k$ (and, for the case of the first integral in Eq. (5.53), above the point $\zeta = -k$). Equation (5.53) is an explicit expression for the current on the infinitely long antenna.

Since the second integral in Eq. (5.53) has (apart from the aforementioned indentation) the form of a Fourier cosine transform, and since the large-ζ behavior of the integrand is known from Eq. (5.48), we can determine the small-z behavior of $I(z)$ using the methods of Section 5.3. We easily see that

$$\frac{I_{\mathrm{ex}}(z)}{V} \sim -i\frac{4ka}{\zeta_0}\ln\frac{1}{|z|}, \quad \text{as}\quad z \to 0. \tag{5.54}$$

This is the leading behavior of the current near the driving point $z = 0$. The logarithmic singularity is expected from physical considerations because there is an infinite capacitance between the two circular knife edges forming the infinitesimal gap at $z = 0$ [9]. Importantly, Eq. (5.54) holds unaltered for the case of the finite antenna of length $2h$ [9] and can be used to accelerate the convergence of moment-method solutions of Hallén's equation [10, 12].

5.5   PROBLEMS

5.1.   Show Eq. (5.14).

5.2.   (i)   Use integration by parts to find the leading term of the large-x asymptotic expansion of the integral $f_2(x, a)$ defined in Eq. (4.6). Compare your answer to Eq. (4.8).

(ii)   Use integration by parts to find the leading term of the large-x asymptotic expansion of the integral $\int_a^b e^{-x\sqrt{t^2+1}}\, dt$. What happens when one of the two integration limits is zero?

5.3.   In Problem 4.8, we encountered the integrals

$$f(x) = \int_0^\infty \frac{\sin t}{t + x}\, dt; \quad g(x) = \int_0^\infty \frac{\cos t}{t + x}\, dt, \quad x > 0 \,.$$

Use integration by parts to find the asymptotic expansions of $f(x)$ and $g(x)$ as $x \to +\infty$.

5.4.   Show Eqs. (5.22) and (5.23). *Hint:* In Eq. (3.1), replace $\int_0^\infty$ by $\int_0^{i\infty}$.

5.5.   Find the first few terms in the asymptotic expansion of $\int_0^\alpha \sqrt{t} \sin(xt)\, dt$ $(\alpha > 0)$ as $x \to +\infty$.

5.6.   For $\alpha > 0$, show that

$$\int_0^\alpha \frac{\sqrt{\alpha - t}}{\alpha^{5/2} + (\alpha - t)^{5/2}} \cos(xt)\, dt \sim \frac{1}{\alpha^{5/2}} \sqrt{\frac{\pi}{8}} \frac{\sin(\alpha x) - \cos(\alpha x)}{x^{3/2}} \quad (x \to +\infty) \,.$$

5.7.   Using a contour integral, show Eq. (5.31).

5.8.   Use integration by parts to explain why

$$\int_0^\infty e^{-\alpha t^2} \cos(xt)\, dt, \quad x > 0, \quad \alpha > 0 \,,$$

is "small" when x is large. Then, verify this smallness by explicitly calculating the integral. (We note that $e^{-\alpha t^2}$ is the standard example of a "rapidly decreasing function" which is important in distribution theory. The Fourier transform of a rapidly decreasing function is known to itself to be a rapidly decreasing function, see [13] for an introductory discussion.)

5.9.   Apply integration by parts to the integral

$$\int_0^\infty \frac{t^3 \sin(xt)}{(1+t^2)^{7/2}} \, dt, \quad x > 0,$$

and discuss the implication of the results.

5.10.   (i) Apply integration by parts to the integral $F_3(x, \alpha)$ defined in Eq. (5.32) and discuss the implications of your findings.

(ii) With the aid of the contour shown in Fig. 5.1 and Jordan's lemma, show that

$$F_3(x, \alpha) = \int_\alpha^\infty \frac{\exp(-xt)}{\sqrt{t^2 - \alpha^2}} \, dt. \tag{5.55}$$

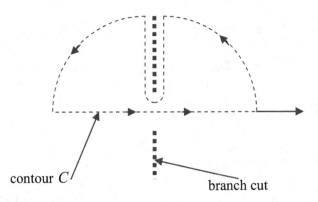

Figure 5.1: Contour for Problem 5.10.

(iii) Apply Watson's lemma to the integral in Eq. (5.55) and thus determine the complete asymptotic expansion of $F_3(x, \alpha)$ $(x \to \infty)$. Is your expansion consistent with your answer in (i)?

(iv) In the usual tables of integrals, we find $F_3(x, \alpha) = K_0(\alpha x)$, where K_0 is the modified Bessel function. Compare (a) your answer in (iii), and (b) the result in Eq. (5.36) to the corresponding results in Appendix A.

5.11.   Using the methods of Section 5.3, show that the large-x behavior of

$$F(x, \alpha) = \int_0^\infty \frac{\cos(xt)}{(t^2 + 1)^{\alpha/2}} \, dt, \quad x > 0, \quad \alpha > 0$$

is given by

$$F(x, \alpha) \sim \begin{cases} \beta x^{\alpha-1}, & 0 < \alpha < 1, \\ \gamma \ln x, & \alpha = 1, \\ O(1), & \alpha > 1. \end{cases}$$

and determine the constants β and γ.

5.12. Using the methods of Section 5.3, show that

$$F(x, x_0) = \int_0^\infty \frac{\sin(xt) \cos\left(x_0\sqrt{t^2+1}\right)}{\sqrt{t^2+1}} \, dt, \qquad x > 0, \quad x_0 > 0$$

is discontinuous at $x = x_0$ and that the jump $F(x_0^+, x_0) - F(x_0^-, x_0)$ at that point is equal to $\pi/2$.

5.13. Using the methods of Section 5.3, show that the behavior of

$$F(x, x_0, \nu) = \int_0^\infty J_\nu(x_0 t) \cos(xt) \, dt, \qquad x > 0, \quad x_0 > 0,$$

when x is close to and larger than x_0, is given by

$$F(x, x_0, \nu) \sim \frac{-\sin(\nu\pi/2)}{\sqrt{2x_0}} \frac{1}{(x - x_0)^{1/2}} \qquad (x \to x_0^+).$$

Find a similar asymptotic formula for the case $x \to x_0^-$.

5.14. (i) Let $K(k)$ be the complete elliptic integral of the first kind (Appendix A). Show that

$$\int_0^{\pi/2} \frac{d\theta}{\sqrt{1 + k^2\sin^2\theta}} = \frac{1}{\sqrt{1 + k^2}} K\left(\frac{k}{\sqrt{1 + k^2}}\right), \qquad 0 < k < 1. \tag{5.56}$$

(ii) Use Eq. (5.56) and a proper asymptotic relation from [1] in order to show Eq. (5.50) directly from Eq. (5.45).

REFERENCES

[1] F. W. J. Olver, D. W. Lozier, R. F. Boisvert, and C. W. Clark, *Digital Library of Mathematical Functions*, National Institute of Standards and Technology from http://dlmf.nist.gov/, §1.8.10,§2.3, §5.9.6, §5.9.7, §7.2.2, §10.40.6. 89, 91, 93, 94, 99, 103

[2] F. W. J. Olver, *Asymptotics and Special Functions*. Natick, MA: A. K. Peters, 1997, Chapter 3. 91, 93

[3] C. M. Bender and S. A. Orszag, *Advanced Mathematical Methods for Scientists and Engineers; Asymptotic Methods and Perturbation Theory*. New York, Springer, 1999, §3.5, §3.8, §6.3. 91, 92

[4] A. Erdélyi, *Asymptotic Expansions*. New York, Dover, 1956. 91, 93

[5] J. P. Boyd, "The Devil's Invention: Asymptotic, Superasymptotic and Hyperasymptotic Series," *Acta Applicandae Mathematicae*, vol. 56, pp. 1–98, 1999. 92

[6] T. M. Apostol, *Mathematical Analysis: A Modern Approach to Advanced Calculus, 2nd ed*. New York, Pearson, 1974. 93

[7] P. D. Miller, *Applied Asymptotic Analysis*. Providence, Rhode Island: American Mathematical Society, 2006. 93

[8] A. P. Prudnikov, Yu. A. Brychkov, and O. I. Marichev, *Integrals and Series: Elementary Functions, vol. 1*. Amsterdam, Gordon & Breach, 1986. 98

[9] T. T. Wu, "Introduction to linear antennas," ch. 8 in *Antenna Theory, pt. I*, R. E. Collin and F. J. Zucker, Eds. New York, McGraw-Hill, 1969. 98, 100

[10] G. Fikioris and T. T. Wu, "On the application of numerical methods to Hallen's equation," *IEEE Trans. Antennas Propagat.*, vol. 49, no. 3, pp. 383–392, March 2001. 98, 100

[11] A. P. Prudnikov, Yu. A. Brychkov, and O. I. Marichev, *Integrals and Series: Elementary Functions, vol. 2*. London: Taylor and Francis, 2002. 99

[12] G. Fikioris, "An Application of Convergence Acceleration Methods," *IEEE Trans. Antennas Propagat.*, vol. 47, no. 12, pp. 1758–1760, Dec. 1999. 100

[13] R. J. Beerends, H. G. ter Morsche, J. C. van den Berg and E. M. van de Vrie, *Fourier and Laplace Transforms*. Cambridge, UK, Cambridge University Press, 2003. 101

CHAPTER 6

Poisson Summation Formula and Applications

The Poisson Summation Formula (PSF) is often used in antenna theory. It transforms certain series (often, these are Fourier series) to other series that frequently converge faster than does the original series. Furthermore, the first few terms of the transformed series may in some sense be a useful asymptotic approximation with respect to a parameter. It may also be the case that the transformed series is a full asymptotic expansion. We discuss several versions of the PSF, including one for finite sums.

6.1 DOUBLY INFINITE SUMS

6.1.1 FORMULA AND ITS DERIVATION

The first version specifically applies to doubly infinite sums, meaning series of the form $\sum_{n=-\infty}^{\infty} f(n)e^{in\theta}$:

Poisson Summation Formula (PSF)

Let $\theta \in \mathbb{R}$ and let $\bar{f}(y) = \int_{-\infty}^{\infty} f(x)e^{iyx}\, dx\ (y \in \mathbb{R})$ denote the Fourier transform of $f(x)$.

Then

$$\sum_{n=-\infty}^{\infty} f(n)e^{in\theta} = \sum_{m=-\infty}^{\infty} \bar{f}(\theta - 2\pi m)\,. \tag{6.1}$$

Our statement of the PSF is not rigorous because we do not give conditions on f under which Eq. (6.1) is valid. The derivation of Eq. (6.1) that follows involves delta functions and is also nonrigorous.

Derivation of PSF

Let $\delta(x)$ denote the Dirac delta function. The equality

$$\sum_{n=-\infty}^{\infty} \delta(x - 2n\pi) = \frac{1}{2\pi} \sum_{m=-\infty}^{\infty} e^{-imx} \tag{6.2}$$

is easily verified using Fourier series: denote by $g(x)$ the function—called impulse train or comb function—in the left-hand side of Eq. (6.2). Since $g(x)$ is 2π-periodic, it has a Fourier-series representation $g(x) = \sum\limits_{m=-\infty}^{\infty} \alpha_m e^{-imx}$ where $\alpha_m = (2\pi)^{-1} \int\limits_{-\pi}^{\pi} g(x)e^{imx}\,dx$. For $g(x) = \sum\limits_{n=-\infty}^{\infty} \delta(x - 2n\pi)$, explicit calculation gives $\alpha_m = 1/(2\pi)$, and Eq. (6.2) follows immediately.

When we multiply the impulse train by a function and then integrate over all x, we obtain a sum whose terms are samples of the function at $2n\pi$, the integer multiples of 2π. Therefore, multiplication of Eq. (6.2) by $f(x/2\pi)e^{ix\theta/2\pi}$ followed by integration from $x = -\infty$ to $x = \infty$ yields an equality whose left-hand side is the same as the left-hand side of Eq. (6.1). Setting $x/(2\pi) = y$ in the integral in the right-hand side, the resulting equality is

$$\sum_{n=-\infty}^{\infty} f(n)e^{in\theta} = \sum_{m=-\infty}^{\infty} \int_{-\infty}^{\infty} f(y)e^{i(\theta-2m\pi)y}\,dy\,. \tag{6.3}$$

With our definition of the Fourier transform, Eq. (6.3) is tantamount to Eq. (6.1).

6.1.2 REMARKS

- In the literature, the name PSF is often used for the formula resulting from Eq. (6.1) by setting $\theta = 0$, viz.,

$$\sum_{n=-\infty}^{\infty} f(n) = \sum_{m=-\infty}^{\infty} \bar{f}(2\pi m)\,. \tag{6.4}$$

Formula (6.4) states that the sum of samples of $f(x)$ at the integers is equal to the sum of samples of its Fourier transform at the integer multiples of 2π. This formula is not really a special case of Eq. (6.1) because Eq. (6.1) easily follows from it (Problem 6.1).

- If $f(x)$ is a non-distributional function, then Eq. (6.1) is a non-distributional formula. Our derivation, however, involved the delta function, which is a distributional function. (This situation is somewhat reminiscent of Green's functions, which are classical functions defined via the delta function.) As one might suspect, there exist classical proofs of Eq. (6.1) (that do not involve delta functions) in the literature: see [1, 2], or the works cited in [3]. We note that different works often have different conditions on f. For a distributional proof of Eq. (6.1) that includes conditions on f, see [4].

- When the function $f(x)$ in Eq. (6.1) is even ($f(x) = f(-x)$, $x \in \mathbb{R}$), we can easily show (Problem 6.2) that Eq. (6.1) amounts to

$$\frac{1}{2}f(0) + \sum_{n=1}^{\infty} f(n)\cos n\theta = \bar{f}_C(\theta) + \sum_{m=1}^{\infty} \left[\bar{f}_C(\theta - 2\pi m) + \bar{f}_C(\theta + 2\pi m)\right]\,, \tag{6.5}$$
$$\theta \in \mathbb{R}\,,$$

where $\bar{f}_C(y) = \int_0^\infty f(x) \cos yx \, dx = \bar{f}_C(-y)$ $(y \in \mathbb{R})$ is the Fourier cosine transform of $f(x)$. The special case of Eq. (6.5) with $\theta = 0$ is

$$\frac{1}{2}f(0) + \sum_{n=1}^\infty f(n) = \bar{f}_C(0) + 2\sum_{m=1}^\infty \bar{f}_C(2\pi m). \qquad (6.6)$$

Formula (6.6) can be found—together with validity conditions and references to works with rigorous proofs—in [3]; we will call it the PSF for semi-infinite sums.

- If (in any of the versions of the PSF) the original sum is slowly convergent, then $f(x)$ is a wide function (similar to a wideband signal). Therefore, its Fourier transform (or its Fourier cosine transform) is a narrow function. For this reason, the transformed sum typically converges faster than does the original one. This significant feature will become apparent in the applications that follow; see also Problem 6.4.

- The PSF has a number of applications in signal processing and communications and readers are likely to have already encountered the PSF (e.g., in [5], where the notation is different from ours). In particular, the important sampling theorem can be viewed as a corollary of the PSF, see Problems 6.5 and 6.6.

- In the applications that follow, we proceed from a sum in the form of the left-hand side of Eq. (6.1) to find the transformed sum on the right-hand side. This procedure involves determining an interpolating function $f(x)$ from its samples $f(n)$ at the integers. Even though $f(x)$ must be such that the transformed sum is convergent—as we have already noted, our statement of the PSF gives no beforehand conditions guaranteeing this—it may be possible to find more than one such function. While different $f(x)$ generally lead to different transformed sums (Problem 6.7), we will not face dilemmas in our applications because $f(n)$ can be extended to the real numbers in a natural manner.

- In order for the transformed sum to be "better" than the original one, a closed-form evaluation of the Fourier transform $\bar{f}(y)$ (or Fourier cosine transform $\bar{f}_C(y)$) is usually desirable. Besides symbolic programs and the usual tables of integrals, a useful work for this purpose is Oberhettinger's table of Fourier transforms [6].

6.1.3   A FIRST EXAMPLE

The sum

$$S(\phi, \tau) = \sum_{n=1}^\infty (-1)^n \frac{\cos(n\phi)}{\cosh(n\tau)}, \quad \tau > 0, \quad 0 \le \phi < \pi \qquad (6.7)$$

pertains to a number of quantities in electrostatic and magnetostatic problems, an example being the DC current distribution at the center $(\rho, \phi, z) = (a, \phi, 0)$ of a conducting cylinder excited by electrodes applied at $(a, 0, \pm h)$, where $\tau = h/a$ [7]. Our first example of using the PSF consists of transforming Eq. (6.7) using Eq. (6.1)—identical results can, of course, be obtained via Eq. (6.5). The first step is to write Eq. (6.7) as

$$S(\phi, \tau) = \frac{1}{2} \left[T(\phi, \tau) - 1 \right] , \tag{6.8}$$

where

$$T(\phi, \tau) = \sum_{n=-\infty}^{\infty} (-1)^n \frac{1}{\cosh(n\tau)} e^{in\phi} , \tag{6.9}$$

which has the form of the left-hand side of Eq. (6.1) with $f(n) = (-1)^n / \cosh(n\tau)$. With $f(x) = \cos(\pi x) / \cosh(\tau x)$ as an interpolating function, Eq. (6.1) gives

$$T(\phi, \tau) = \sum_{m=-\infty}^{\infty} \int_{-\infty}^{\infty} e^{i(\phi - 2\pi m)x} \frac{\cos(\pi x)}{\cosh(\tau x)} \, dx . \tag{6.10}$$

Equation (6.10) can be written as

$$T(\phi, \tau) = \frac{1}{2} \sum_{m=-\infty}^{\infty} \int_{-\infty}^{\infty} \frac{e^{i[\phi - (2m-1)\pi]x}}{\cosh(\tau x)} \, dx + \frac{1}{2} \sum_{m=-\infty}^{\infty} \int_{-\infty}^{\infty} \frac{e^{i[\phi - (2m+1)\pi]x}}{\cosh(\tau x)} \, dx . \tag{6.11}$$

The two series in Eq. (6.11) are identical, so that

$$T(\phi, \tau) = \sum_{m=-\infty}^{\infty} \int_{-\infty}^{\infty} \frac{e^{i[\phi - (2m+1)\pi]x}}{\cosh(\tau x)} \, dx = 2 \sum_{m=-\infty}^{\infty} \int_{0}^{\infty} \frac{\cos\left\{[\phi - (2m+1)\pi]x\right\}}{\cosh(\tau x)} \, dx . \tag{6.12}$$

The integral in (6.12) is a Fourier cosine transform that can be found in [6] (also, in tables of integrals), evaluated by symbolic programs, or found via a contour integral (Problem 6.8):

$$\int_{0}^{\infty} \frac{\cos(yx)}{\cosh(ax)} \, dx = \frac{\pi}{2a \cosh[\pi y/(2a)]}, \quad a > 0, \quad y \in \mathbb{R} . \tag{6.13}$$

Combining Eq. (6.8), (6.12), and Eq. (6.13), we obtain our final result

$$S(\phi, \tau) = -\frac{1}{2} + \frac{\pi}{2\tau} \sum_{m=-\infty}^{\infty} \frac{1}{\cosh \frac{[(2m+1)\pi - \phi]\pi}{2\tau}} . \tag{6.14}$$

In the special case $\phi = 0$, Eq. (6.14) can be written as a semi-infinite sum,

$$S(0, \tau) = -\frac{1}{2} + \frac{\pi}{\tau} \sum_{m=0}^{\infty} \frac{1}{\cosh \frac{(2m+1)\pi^2}{2\tau}} . \tag{6.15}$$

Since $\cosh x \sim e^x/2$ $(x \to +\infty)$, the series in Eq. (6.14) and Eq. (6.15) are both convergent. While the original series Eq. (6.7) contains both positive and negative terms (such series are prone to cancellation errors), all terms of the transformed series Eqs. (6.14) and (6.15) are positive. The convergence properties of the original and the transformed series are complementary: when τ increases, the convergence properties of the original series improve, while those of the transformed series worsen. Precisely the opposite occurs when τ decreases.

The numerical advantages of the transformed series thus occur for small τ. The advantages can be very significant: when $\tau = 1$, retaining $-1/2$ together with a single term in the series of Eq. (6.15) gives an accuracy of 0.0005%, whereas Eq. (6.7) achieves a comparable accuracy only after 12-13 terms. Equation (6.15) is even better for smaller values of τ.

From the point of view of asymptotics, Eq. (6.15) is a (convergent) asymptotic expansion as $\tau \to 0$. The relevant asymptotic sequence (see Section 2.4) is $\psi_{-1}(\tau)$, $\psi_0(\tau)$, $\psi_1(\tau), \ldots$, where

$$\psi_{-1}(\tau) = -\frac{1}{2}, \quad \psi_m(\tau) = \frac{\pi}{\tau} \frac{1}{\cosh \frac{(2m+1)\pi^2}{2\tau}} \quad (m = 0, 1, 2, \ldots) . \tag{6.16}$$

The normalizations in Eq. (6.16) are such that all coefficients of $\psi_m(\tau)$ in Eq. (6.15) are 1. The transformed series Eq. (6.15) (that was generated by the PSF) can thus be considered an asymptotic expansion—associated with a rather unusual asymptotic sequence—for the case where the parameter τ is small. When $\phi \neq 0$, the situation with the series Eq. (6.14) is not very different: the elements of the relevant asymptotic sequence, which now also depend on ϕ, are

$$\xi(\phi, \tau) = -\frac{1}{2}, \quad \xi_m(\phi, \tau) = \frac{\pi}{2\tau} \frac{1}{\cosh \frac{[(2m+1)\pi - \phi]\pi}{2\tau}} \quad (m \in \mathbb{Z}) . \tag{6.17}$$

Since $\cosh x$ is even and increases for $x \geq 0$, the correct ordering of the sequence elements in Eq. (6.17) is the same as the ordering of $|(2m+1)\pi - \phi|$. This ordering can be understood from Fig. 6.1, where we plot $|(2m+1)\pi - \phi|$ as a function of ϕ $(0 < \phi < \pi)$ for various m. It is

$$\xi(\phi, \tau), \quad \xi_0(\phi, \tau), \quad \xi_{-1}(\phi, \tau), \quad \xi_1(\phi, \tau), \quad \xi_{-2}(\phi, \tau), \quad \xi_2(\phi, \tau), \ldots . \tag{6.18}$$

The ordering in Eq. (6.18) means that $\xi(\phi, \tau) \gg \xi_0(\phi, \tau) \gg \xi_{-1}(\phi, \tau) \gg \ldots$ as $\tau \to 0$. This ordering is only correct when $0 < \phi < \pi$: when $\phi = 0$, it is apparent from Fig. 6.1 that $\xi_0(0, \tau) = \xi_{-1}(0, \tau)$, $\xi_1(0, \tau) = \xi_{-2}(0, \tau)$, etc. In practical terms this means that, while valid, the asymptotic approximation

$$S(\phi, \tau) \sim \xi(\phi, \tau) + \xi_0(\phi, \tau) \quad (\tau \to 0) \quad (0 < \phi < \pi) \tag{6.19}$$

deteriorates in numerical accuracy when ϕ is small. On the other hand, retaining one more term, as in

$$S(\phi, \tau) \sim \xi(\phi, \tau) + \xi_0(\phi, \tau) + \xi_{-1}(\phi, \tau) \quad (\tau \to 0) \quad (0 \leq \phi < \pi) , \tag{6.20}$$

Figure 6.1: Plot of $|(2m + 1)\pi - \phi|$ as a function of ϕ for $m = 0, -1, +1, -2, +2$.

gives an approximation that is good even for small ϕ. Only when ϕ is small does the third term become significant. In the extreme case $\phi = 0$, the third term becomes equal to the second term and Eq. (6.20) amounts to retaining $-1/2$ together with a single term in the series in Eq. (6.15) (namely, the $m = 0$ term). A reasoning like this is often necessary when applying the PSF: see, for example, Sections 6.2.5 and 8.2 below. In Problem 6.11, the reader is asked to numerically demonstrate the superiority of Eq. (6.20) over Eq. (6.19).

6.1.4   APPLICATION: INFINITE LINEAR ARRAY OF TRAVELING-WAVE CURRENTS

Properly dimensioned, large circular arrays of cylindrical dipoles with only one or two dipoles driven are known to possess very narrow resonances [8, 9]. At each resonance, the currents on the array elements can be thought of as samples of a slow traveling wave (case of two driven elements) or of a slow standing wave (case of one driven element). The purpose of [8] is to study the near fields of such arrays. To achieve this, [8] first analyzes the near fields of the following simplified continuous or discrete two-dimensional (2-D) problems,

   (i) the infinite, planar sheet of traveling-wave current (continuous),

   (ii) the infinite linear array of traveling-wave currents (discrete),

(iii) the infinite circular cylinder of traveling-wave current (continuous),

(iv) the infinite circular array of N traveling-wave currents (discrete).

The sources in the two continuous problems (i) and (iii) are z-directed surface current densities $\mathbf{K}(\mathbf{r}') = \hat{\mathbf{z}} K_z(\mathbf{r}')$ (ampere per meter) distributed on a 2-D surface with cross-section C. The resulting electric fields are z-directed, $\mathbf{E}(\mathbf{r}) = \hat{\mathbf{z}} E_z(\mathbf{r})$. Assuming an $\exp(-i\omega t)$ time dependence ($\omega = 2\pi f = ck = 2\pi c/\lambda$, with c the velocity of light in free space and k the free-space wavenumber), $E_z(\mathbf{r})$ can be found from the line integral [8]

$$E_z(\mathbf{r}) = -\frac{k\zeta_0}{4} \int_C K_z(\mathbf{r}') H_0^{(1)}\left(k\,|\mathbf{r} - \mathbf{r}'|\right)\, dl\,, \tag{6.21}$$

where ζ_0 is the free-space impedance. Equation (6.21) involves the 2-D Green's function $H_0^{(1)}\left(k\,|\mathbf{r} - \mathbf{r}'|\right)$, where $H_0^{(1)}$ is the Hankel function (Appendix A). In Eq. (6.21), $|\mathbf{r} - \mathbf{r}'|$ is the distance from the observation point $\mathbf{r}$ to a source point $\mathbf{r}'$ on the surface C.

   Problems (ii) and (iv) involve discrete, z-directed current filaments I_n (ampere). If these are located at $\mathbf{r}'_n$, the electric field $\mathbf{E}(\mathbf{r})$ equals $\hat{\mathbf{z}} E_z(\mathbf{r})$, where

$$E_z(\mathbf{r}) = -\frac{k\zeta_0}{4} \sum_n I_n H_0^{(1)}\left(k\,|\mathbf{r} - \mathbf{r}'_n|\right)\,. \tag{6.22}$$

The currents in problem (ii) are samples of the surface current densities in problem (i) and, as will be seen shortly, connections between the respective fields can be brought out by the PSF (6.1). Similarly, the N currents in problem (iv) are samples of the surface current densities in problem (iii); the mathematical tool that can connect the two problems is the PSF for finite sums, to be discussed in Section 6.2.1—problems (iii) and (iv) will themselves be discussed in Section 6.2.5.

   We proceed with (i): Our infinite, planar sheet of traveling-wave current is located on the $y = 0$ plane and given by

$$K_z^P(x') = K_z^P(0) e^{i\beta x'}, \quad -\infty < x' < \infty\,, \tag{6.23}$$

where the wavenumber β of the traveling wave is assumed to be real. The sheet extends to infinity in both the z- and x-directions. When $|\beta| < k$ ($|\beta| > k$), the phase velocity $\omega/|\beta|$ is greater than c (smaller than c) so that the "current wave" given by Eq. (6.23) is fast (slow). Using cartesian coordinates, Eq. (6.21) becomes

$$E_z(x, y) = -\frac{k\zeta_0 K_z^P(0)}{4} \int_{-\infty}^{\infty} e^{i\beta x'} H_0^{(1)}\left(k\sqrt{(x - x')^2 + y^2}\right) dx'\,. \tag{6.24}$$

The integral in Eq. (6.24) can be evaluated analytically with the aid of [6], and one obtains

$$
E_z(x,y) = \begin{cases} \dfrac{-k\zeta_0 K_z^P(0)}{2\sqrt{k^2-\beta^2}} e^{i\beta x} e^{i|y|\sqrt{k^2-\beta^2}} , & |\beta| < k , \\[4mm] \dfrac{ik\zeta_0 K_z^P(0)}{2\sqrt{\beta^2-k^2}} e^{i\beta x} e^{-|y|\sqrt{\beta^2-k^2}} , & |\beta| > k . \end{cases}
\tag{6.25}
$$

This field is symmetric (i.e., even in y) as expected. The derivative $\partial E_z/\partial y$ is discontinuous at $y = 0$, with the discontinuity proportional to $K_z^P(x')$. In the fast-wave case (top line in Eq. (6.25)) the field on either side of the current sheet is a plane wave, traveling outward from the sheet toward the direction $\pm\hat{\mathbf{x}}\beta \pm \hat{\mathbf{y}}\sqrt{k^2-\beta^2}$, with the upper sign corresponding to $y > 0$. In the slow-wave case (bottom line in Eq. (6.25)), the field is an inhomogeneous plane wave and, more precisely, a surface wave, just as that occurring in the well-known phenomenon of total internal reflection [10]: the field propagates parallel to the sheet in the x-direction, but decays exponentially perpendicular to the sheet in the y-direction. Note that the field Eq. (6.25) does not satisfy the 2-D radiation condition; this is logical because the current sheet extends to infinity in the x-direction.

We now move to problem (ii) and its connections to (i). If we sample the surface current densities in Eq. (6.23) at the equispaced points $(x,y) = (nd,0)$ ($n \in \mathbb{Z}$), we obtain an infinite number of z-directed line currents I_n^P given by

$$
I_n^P = I_0^P e^{i\beta dn}, \quad n \in \mathbb{Z} .
\tag{6.26}
$$

The I_n^P form a 2-D, infinite linear array of traveling-wave currents. Adding multiples of 2π to βd leaves the currents unaltered. Thus, with no loss of generality for this discrete problem, we can assume that

$$
-\pi < \beta d < \pi ,
\tag{6.27}
$$

where we exclude $\beta d = \pi$ for simplicity. Equations (6.22) and (6.26) give

$$
E_z(x,y) = -\frac{k\zeta_0 I_0^P}{4} \sum_{n=-\infty}^{\infty} e^{i\beta dn} H_0^{(1)}\left(k\sqrt{(x-nd)^2+y^2}\right) .
\tag{6.28}
$$

From the large-argument asymptotic approximation for the Hankel function (Appendix A), we see that the n-th term of the sum in Eq. (6.28) behaves like $i^{-|n|} e^{i(\beta+k)d|n|}/\sqrt{|n|}$ for large $|n|$. Thus, the infinite sum converges slowly and is not useful for numerical evaluation. (Such slow convergence can often motivate us to utilize the PSF.) Applying Eq. (6.1) to Eq. (6.28), we obtain

$$
E_z(x,y) = \frac{-k\zeta_0 I_0^P}{4d} \sum_{p=-\infty}^{+\infty} e^{iB_p x} \int_{-\infty}^{+\infty} e^{iB_p t} H_0^{(1)}\left(k\sqrt{t^2+y^2}\right) dt ,
\tag{6.29}
$$

where

$$
B_p = \beta + \frac{2\pi p}{d} .
\tag{6.30}
$$

The integral in Eq. (6.29) was encountered previously in Eq. (6.24) so, similar to Eq. (6.25), we obtain

$$
E_z(x, y) = \frac{-k\zeta_0 I_0^P}{2d} \sum_{p=-\infty}^{+\infty} e^{iB_p x}
\begin{cases}
\dfrac{1}{\sqrt{k^2 - B_p^2}} e^{i|y|\sqrt{k^2 - B_p^2}} \,, & |B_p| < k \,, \\[4mm]
\dfrac{-i}{\sqrt{B_p^2 - k^2}} e^{-|y|\sqrt{B_p^2 - k^2}} \,, & |B_p| > k \,.
\end{cases}
\tag{6.31}
$$

The study in [8] pertains to the resonant circular arrays of [9]; in such arrays, the interelement spacing is smaller than $\lambda/2$, or

$$
kd < \pi \,,
\tag{6.32}
$$

an assumption that is also made herein and in [8]. Subject to Eq. (6.32), the first condition $|B_p| < k$ in Eq. (6.31) can only occur when $p = 0$. Our final result thus is

$$
E_z(x, y) = e_0(x, y) + \frac{ik\zeta_0 I_0^P}{2d} \sum_{p \neq 0} e^{iB_p x} \frac{1}{\sqrt{B_p^2 - k^2}} e^{-|y|\sqrt{B_p^2 - k^2}} \,,
\tag{6.33}
$$

where

$$
e_0(x, y) =
\begin{cases}
\dfrac{-k\zeta_0 I_0^P}{2d\sqrt{k^2 - \beta^2}} e^{i\beta x} e^{i|y|\sqrt{k^2 - \beta^2}} \,, & |\beta| < k \,, \\[4mm]
\dfrac{ik\zeta_0 I_0^P}{2d\sqrt{\beta^2 - k^2}} e^{i\beta x} e^{-|y|\sqrt{\beta^2 - k^2}} \,, & k < |\beta| < \dfrac{\pi}{d} \,.
\end{cases}
\tag{6.34}
$$

Comparison of Eq. (6.34) with Eq. (6.25) shows that $e_0(x, y)$ is simply the field due to a z-directed, infinite current sheet, with the top (bottom) line in Eq. (6.34) corresponding to a fast- (slow-) wave. The said current sheet is given by Eq. (6.23), with β the same as in our discrete problem and with $K_z^P(0) = I_0^P/d$. Unless $|y|$ is very small and we are very close to the discrete currents, the series in Eq. (6.33) converges very rapidly. When $y = 0$ and $x \neq nd$ the series does converge, but slowly. At points $(x, y) = (nd, 0)$, the sum diverges logarithmically, due to the presence of 2-D sources at these points.

Besides providing a series that converges faster than the original one, the PSF has established that the discrete problem is, at large observation distances $|y|$, close to the continuous one. This closeness can be conveniently cast in the form of the asymptotic approximation

$$
E_z(x, y) \sim e_0(x, y) \quad (|y| \to \infty) \,,
\tag{6.35}
$$

to which the sum in Eq. (6.33) can be considered as the remainder. The reader is asked to show Eq. (6.35) in Problem 6.12.

Many of the characteristics brought out by the PSF are illustrated in the 3-D plot of the electric field provided in Fig. 6.2. Fig. 6.2 has been calculated from Eq. (6.33) (with Eqs. (6.34)

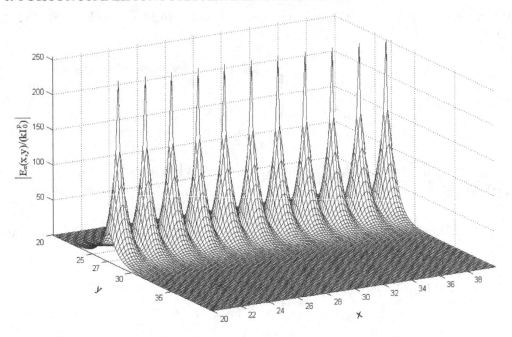

Figure 6.2: Magnitude $|E_z(x,y)/(kI_0^P)|$ of an electric field due to an infinite linear array of traveling-wave currents; $kd = 1.885$ and $\beta d = 2.793$. *(Figure from Fikioris and Matos [8].)*

and (6.30)) using a sufficiently large number of terms (very few terms were necessary, except for very small $k|y|$), and the infinities at $(x, y) = (nd, 0)$ have been truncated. The rapid decrease of the field is apparent. In both problems (i) and (ii), the decrease is exponential (except for very small $k|y|$), and continues indefinitely; if one uses a logarithmic scale on the vertical axis, the decrease appears to be linear.

6.1.5 APPLICATION: COUPLED PSEUDOPOTENTIAL ARRAYS

We will be much briefer in our treatment of a sum that is, in a sense, the 3-D analog of that in Eq. (6.28). This sum arises in [11], a work dealing with traveling waves in two coupled, parallel, infinite, linear point-scatterer (or pseudopotential) arrays. (It also arises in a different context in [12].) With a $e^{j\omega t}$ time dependence, let $G_0(r) = e^{-jk_0 r}/(4\pi r)$ be the 3-D Green's function of the Helmholtz operator $\nabla^2 + k_0^2$. The sum in question is

$$S = \sum_{n=-\infty}^{+\infty} G_0\left(\sqrt{(nb_x - b_0)^2 + b_y^2}\right) e^{-jn\theta},$$

(6.36)

where θ is the Bloch phase shift, i.e., the phase difference between elements in successive unit cells, b_x is the interelement spacing (which is the same in the two arrays), b_0 the offset along the direction parallel to the arrays, and b_y is the perpendicular distance between the two arrays. Application of Eq. (6.1) gives

$$S = \sum_{m=-\infty}^{+\infty} \int_{-\infty}^{+\infty} e^{j(2m\pi-\theta)x} G_0 \left(\sqrt{(xb_x - b_0)^2 + b_y^2} \right) dx \,. \tag{6.37}$$

Setting $t = k_0 (xb_x - b_0)$ we obtain

$$S = \frac{1}{2\pi b_x} \sum_{p=-\infty}^{+\infty} e^{j(2m\pi-\theta)\frac{b_0}{b_x}} \int_{-\infty}^{+\infty} \cos\left(\frac{2m\pi - \theta}{k_0 b_x} t \right) \frac{e^{-j\sqrt{t^2+(k_0 b_y)^2}}}{\sqrt{t^2 + (k_0 b_y)^2}} \, dt \,. \tag{6.38}$$

The integrals are provided in [6], leading to the final formula

$$S = \frac{1}{2\pi b_x} \sum_{m=-\infty}^{+\infty} e^{j(2m\pi-\theta)\frac{b_0}{b_x}}$$

$$\times \begin{cases} -j\frac{\pi}{2} H_0^{(2)} \left(\frac{b_y}{b_x} \sqrt{(k_0 b_x)^2 - (\theta - 2m\pi)^2} \right), & |\theta - 2m\pi| < k_0 b_x \\[2mm] K_0 \left(\frac{b_y}{b_x} \sqrt{(\theta - 2m\pi)^2 - (k_0 b_x)^2} \right), & |\theta - 2m\pi| > k_0 b_x \,, \end{cases} \tag{6.39}$$

where $H_0^{(2)}$ and K_0 are the Hankel and modified Bessel functions, discussed in Appendix A. Note that the first condition is never satisfied when the slow-wave condition $k_0 b_x < |\theta| \leq \pi$ is satisfied. Note also the qualitative similarities of Eq. (6.39) to (6.33). The implications of Eq. (6.39) are fully discussed in [11] (see also [12]).

6.2   FINITE SUMS

We now discuss a version of the Poisson Summation Formula appropriate for finite sums, abbreviated PSF-FS.

6.2.1   FORMULA AND PROOF

Poisson Summation Formula for Finite Sums (PSF-FS)
    Let $g(x)$ be a continuous function defined on the closed interval $[0, C]$ and let $N = 1, 2, \ldots$. Then

$$\sum_{n=0}^{N} g\left(n\frac{C}{N} \right) = \frac{1}{2}g(0) + \frac{1}{2}g(C) + N \sum_{m=-\infty}^{\infty} \alpha_{mN} \,, \tag{6.40}$$

where

$$\alpha_q = \frac{1}{C} \int\limits_0^C g(x) e^{-i2\pi qx/C} \, dx, \quad q \in \mathbb{Z} \tag{6.41}$$

are the Fourier-series coefficients of $g(x)$. A form obviously equivalent to Eq. (6.40) is

$$\frac{1}{2}g(0) + \sum_{n=1}^{N-1} g\left(n\frac{C}{N}\right) + \frac{1}{2}g(C) = N \sum_{m=-\infty}^{\infty} \alpha_{mN} \,. \tag{6.42}$$

In the equations above and throughout the remainder of this section, the series $\sum\limits_{m=-\infty}^{\infty}$ is to be interpreted as $\lim\limits_{M\to\infty} \sum\limits_{m=-M}^{M}$ (this is usually called the principal value of the series—it is the limit of the symmetric partial sums).

A derivation of the PSF-FS is possible (Problem 6.13; see also [13], [14]) using the PSF for semi-infinite sums, Eq. (6.6). That derivation, however, involves extending the definition of $g(x)$ beyond the interval $[0, C]$ and, in particular, to $x = +\infty$. In what follows, we present a different derivation that does not require such an extension. The derivation is taken from Section IV.A of [15], a section that is, in turn, based upon ideas that can be found in [16].

Proof of PSF-FS

The *simple trapezoidal rule* is familiar from elementary numerical analysis: one approximates the integral $I(g) = \int\limits_0^C g(x)\, dx$ by the quantity $T_1(g) = C\left[g(0)/2 + g(C)/2\right]$. More general is the *trapezoidal rule* (often called composite trapezoidal rule), where the approximation $T_N(g)$ $(N = 1, 2, \ldots)$ of $I(g)$ results from the following procedure: divide the integration interval $(0, C)$ into N subintervals, each of length C/N, approximate each of the resulting sub-integrals by the simple trapezoidal rule, and sum to obtain $T_N(g)$. In the course of this procedure, all samples of $g(x)$ occur twice, except for the endpoint samples $g(0)$ and $g(C)$, which only occur once. The trapezoidal rule is thus

$$I(g) \equiv \int\limits_0^C g(x)\, dx = T_N(g) + E_N(g) \,, \tag{6.43}$$

where $E_N(g)$ is the error of the trapezoidal rule and

$$T_N(g) = \frac{C}{N}\left[\frac{1}{2}g(0) + \sum_{n=1}^{N-1} g\left(n\frac{C}{N}\right) + \frac{1}{2}g(C)\right]. \tag{6.44}$$

Consider the Fourier series of $g(x)$,

$$\sum_{q=-\infty}^{\infty} \alpha_q e^{i2\pi qx/C} = \alpha_0 + \sum_{q=1}^{\infty} \left(\alpha_q e^{i2\pi qx/C} + \alpha_{-q} e^{-i2\pi qx/C}\right), \quad x \in \mathbb{R}, \tag{6.45}$$

with the coefficients α_q given by Eq. (6.41). For all real x, the periodic Fourier series in Eq. (6.45) converges to (is equal to) the *periodic extension* of $g(x)$ at points of continuity, and to the mean value at points of finite discontinuity. (A useful, simple discussion on the convergence of Fourier series is contained in [17].) For example, the periodic extension of $g(x) = x$, $(x \in [0, C])$ is a sawtooth function with minimum (maximum) values 0 (C); from Eq. (6.41) and the second form Eq. (6.45) it is seen that the Fourier series of $g(x)$ is $C/2 - (C/\pi) \sum_{q=1}^{\infty} \sin(2\pi qx/C)/q$, which, as expected, equals $C/2$ when $x = 0$ or $x = C$.

Therefore, for $x \in [0, C]$ we have

$$
\sum_{q=-\infty}^{\infty} \alpha_q e^{i2\pi qx/C} = \begin{cases} g(x), & \text{if} \quad x \in (0, C) , \\ \dfrac{1}{2}g(0) + \dfrac{1}{2}g(C), & \text{if} \quad x = 0 \quad \text{or} \quad x = C . \end{cases} \tag{6.46}
$$

Now express $I(g)$ and $T_N(g)$ in terms of the Fourier coefficients α_q. From Eq. (6.41),

$$
I(g) \equiv \int_0^C g(x)\, dx = C\alpha_0 . \tag{6.47}
$$

In Eq. (6.44), substitute the samples of $g(x)$ using Eq. (6.46) to obtain

$$
T_N(g) = \frac{C}{N} \sum_{n=0}^{N-1} \sum_{q=-\infty}^{\infty} \alpha_q e^{i2\pi qn/N} . \tag{6.48}
$$

Now interchange the order of summation. The resulting summation over n equals N when $q = mN$ ($m \in \mathbb{Z}$), and vanishes otherwise. Therefore,

$$
T_N(g) = C \sum_{m=-\infty}^{\infty} \alpha_{mN} = C \left[\alpha_0 + \sum_{m=1}^{\infty} (\alpha_{mN} + \alpha_{-mN}) \right] . \tag{6.49}
$$

Together, formulas (6.43), (6.44), (6.47), and (6.49) are

$$
\int_0^C g(x)\, dx = \frac{C}{N} \left[\frac{1}{2}g(0) + \sum_{n=1}^{N-1} g\left(n\frac{C}{N}\right) + \frac{1}{2}g(C) \right] - C \sum_{m=1}^{\infty} (\alpha_{mN} + \alpha_{-mN}) . \tag{6.50}
$$

Rearrangement gives Eq. (6.42), from which Eq. (6.40) follows immediately.

6.2.2   REMARKS

- Besides not relying on the PSF for semi-infinite sums, our derivation of PSF-FS proceeds from first principles and clearly shows the origin of the extra terms involving $g(0)$ and

$g(C)$. Furthermore, it is classical in the sense that it makes no use of distributions such as delta functions. According to [18], the PSF-FS has been used improperly in the past in the antennas/electromagnetics literature. In this regard, we note that our $g(x)$ is defined exclusively in the interval $[0,C]$, so that $g(0)$ and $g(C)$ in Eqs. (6.40) and (6.42) necessarily mean $\lim_{x\to 0^+} g(x)$ and $\lim_{x\to C^-} g(x)$, respectively, in accordance with the notation of [18].

- The PSF-FS is usually viewed as an alternative representation for the finite sum on the left-hand side of Eq. (6.40) or (6.42). Because of Eqs. (6.47) and (6.49), the PSF-FS can also be interpreted as an expression for the error $E_N(g)$ in the trapezoidal rule; the said expression involves $\alpha_{\pm N}, \alpha_{\pm 2N}, \ldots$, where α_q is the qth Fourier coefficient of the integrand $g(x)$.

6.2.3   ELEMENTARY EXAMPLE

Problem 6.14 asks the reader to verify the PSF-FS for two elementary examples for which the corresponding semi-infinite sum $\sum_{n=0}^{\infty} g(nC/N)$ diverges (so that the steps followed in the derivation of Problem 6.13 are not directly applicable). Here, we do the same for $g(n) = n$, which is also associated with a divergent semi-infinite sum. As an interpolating function, we choose $g(x) = x$. The left-hand side of Eq. (6.40) equals $C(N+1)/2$ and the right-hand side equals $C/2 + N \sum_{m=-\infty}^{\infty} \alpha_{mN}$. By elementary manipulations, we find that Eq. (6.41) gives $\alpha_0 = C/2$ and $\alpha_q = iC/(2\pi q)$ for $q = \pm 1, \pm 2, \ldots$. Thus, $\alpha_q + \alpha_{-q} = 0$ for $q = \pm 1, \pm 2, \ldots$, implying that the symmetric partial sum $\sum_{m=-M}^{M} \alpha_{mN}$ equals α_0, or $C/2$. Therefore, the principal value of $\sum_{m=-\infty}^{\infty} \alpha_{mN}$ also equals $C/2$, thus verifying Eq. (6.40). Note that interpretation as a principal value was necessary, as the double limit $\lim_{M\to\infty} \lim_{M'\to\infty} \sum_{m=-M'}^{M} \alpha_{mN}$ diverges.

6.2.4   CONTINUOUS FUNCTIONS WITH EQUAL ENDPOINT VALUES

Suppose now that $g(0) = g(C)$, so that that the endpoint values of the continuous function $g(x)$ are equal and $g(x)$ has a *continuous periodic extension*. Formulas (6.40) and (6.42) become simply

$$\frac{C}{N} \sum_{n=0}^{N-1} g\left(n\frac{C}{N}\right) = \sum_{m=-\infty}^{\infty} \int_0^C g(x) e^{-i2\pi mNx/C} \, dx \quad \text{(provided that } g(0) = g(C)\text{)}, \quad (6.51)$$

where Eq. (6.41) was used. The finite sum on the left-hand side of Eq. (6.51) has been written as an infinite sum of integrals and, in this special case, no extra terms appear. By Eq. (6.44), the sum on the left-hand side of Eq. (6.51) is the trapezoidal-sum approximation $T_N(g)$ for the $m = 0$ term (integral) on the right-hand side of Eq. (6.51). In this special case, $T_N(g)$ consists of N samples of $g(x)$, each appearing exactly once; the endpoint value $g(C)$ does not appear.

An equivalent to Eq. (6.51) form is

$$\frac{C}{N}\sum_{n=0}^{N-1} g\left(n\frac{C}{N}\right) = \int_0^C g(x)\,dx + \sum_{m\neq 0}\int_0^C g(x)e^{-i2\pi mNx/C}\,dx \tag{6.52}$$

(provided that $g(0) = g(C)$) .

The sum on the right-hand side of Eq. (6.52) is the error $E_N(g)$ of the trapezoidal rule. Under suitable conditions on the periodic extension of $g(x)$, this error can be estimated. As it turns out (precise theorems for real-valued functions can be found in [16, 19]) the trapezoidal rule is, for periodic functions, "superior to any other quadrature rule" [19]; better for *smoother* periodic functions; and for *analytic* periodic functions, the error is exponentially small in the number N of samples. In other words, under such conditions, the first term (integral) of the RHS of Eq. (6.52) is an excellent large-N approximation to the finite sum on the left-hand side.

6.2.5   APPLICATION: CYLINDRICAL ARRAY OF TRAVELING-WAVE CURRENTS

In the present section, we apply the PSF-FS to the sum associated with problem (iv) of Section 6.1.4 [8]. We first discuss problem (iii)—as already mentioned, the PSF-FS will establish a connection between the two problems. The z-directed, traveling-wave source of problem (iii) is a continuous cylindrical current sheet (surface current density) of radius a, extending to infinity in the z-direction and given by

$$K_z^C(\phi') = K_z^C(0)e^{im\phi'}, \quad 0 \le \phi' \le 2\pi, \quad m \in \mathbb{Z} . \tag{6.53}$$

For the current wave in Eq. (6.53), one can define an associated wavenumber β_m, a wavelength λ_m, and a phase velocity v_m by

$$\beta_m = \frac{m}{a}, \qquad \lambda_m = \frac{2\pi}{|\beta_m|}, \qquad v_m = \frac{2\pi f}{\beta_m} . \tag{6.54}$$

The circumference $2\pi a$ of the cylinder is always an integer multiple of wavelengths λ_m. There are several ways to write the condition for a slow wave, e.g.,

$$|v_m| < c \quad \text{or} \quad k < |\beta_m| \quad \text{or} \quad ka < |m| . \tag{6.55}$$

Using polar coordinates (ρ, ϕ), Eq. (6.21) becomes

$$E_z(\rho, \phi) = -\frac{ka\zeta_0 K_z^C(0)}{4}\int_0^{2\pi} e^{im\phi'} H_0^{(1)}\left(k\sqrt{\rho^2 + a^2 - 2\rho a \cos(\phi - \phi')}\right) d\phi' . \tag{6.56}$$

To evaluate the integral in Eq. (6.56), set $\phi - \phi' = \theta$. The resulting integral is 2π times the m-th Fourier-series coefficient of $H_0^{(1)}\left(k\sqrt{\rho^2 + a^2 - 2\rho a \cos\theta}\right)$, a coefficient which is known from the addition theorem (A.44) of Appendix A. We thus obtain

$$E_z(\rho,\phi) = \begin{cases} \dfrac{-ka\pi\zeta_0 K_z^C(0)}{2} e^{im\phi} H_m^{(1)}(ka) J_m(k\rho) \ , & \rho < a \ , \\[2em] \dfrac{-ka\pi\zeta_0 K_z^C(0)}{2} e^{im\phi} J_m(ka) H_m^{(1)}(k\rho) \ , & \rho > a \ . \end{cases} \tag{6.57}$$

Here, the field at $a + \rho$ is not equal to the field at $a - \rho$ (compare to the symmetry of problem (i) discussed in Section 6.1.4). By the small-argument approximation Eq. (A.23) of J_m, the interior field has a zero of order m at $\rho = 0$, so that the field vanishes very rapidly as one approaches the origin. These characteristics are illustrated in Fig. 6.3. Note that the scale on the

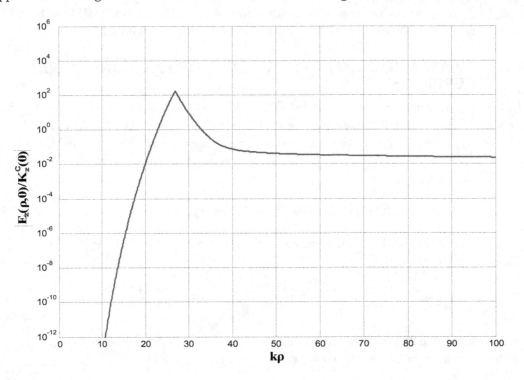

Figure 6.3: Magnitude $|E_z(\rho,0)/K_z^C(0)|$ of an electric field due to a cylinder of traveling-wave current, as function of $k\rho$; $ka = 27$ and $m = 40$. The scale on the vertical axis is logarithmic. *(Figure from Fikioris and Matos [8].)*

vertical axis is logarithmic. As one moves away from the cylinder towards infinity, it is apparent that, initially, the field decreases very rapidly. The rate of decrease then slows down. For sufficiently large ρ, and by Eq. (A.34), E_z decreases as $\exp\left(ik\rho\right)/\sqrt{\rho}$ in accordance with the radiation condition (as opposed to the planar case, in which the rapid—exponential—decrease continues indefinitely and the radiation condition does not hold, see Eq. (6.25)). As one moves *inwards* from $\rho = a$, the decrease is even more rapid; here, the decrease continues until one reaches the origin.

When ka is a zero of the Bessel function, i.e., $J_m(ka) = 0$ (this is not the case in Fig. 6.3), it is seen from Eq. (6.57) that the field exterior to the cylinder vanishes identically for all ρ and ϕ. In such exceptional cases, we have "nonradiating currents." These have been extensively discussed in the literature [20, 23] and in fact, our 2-D cylindrical case is probably the simplest example in which nonradiating currents occur. By an inequality in [3], the condition $J_m(ka) = 0$ can only occur when $m < ka$, so that Eq. (6.55) shows that nonradiating currents cannot occur when the current wave is slow. The field interior to the cylinder, on the other hand, can never vanish identically; this is because $H_m^{(1)}$ has no real zeros [3].

Sampling the surface current density in Eq. (6.53) gives the 2-D circular array of traveling-wave currents, problem (iv). If there are N z-directed line currents I_n^C, located at $(\rho, \phi) = (a, 2\pi n/N)$ $(n = 0, 1, \ldots, N - 1)$, one has

$$I_n^C = I_0^C e^{i2\pi mn/N}, \quad n = 0, 1, \ldots, N - 1 . \tag{6.58}$$

Adding multiples of N to m does not change the currents, so we can assume that

$$- N/2 < m \leq N/2 . \tag{6.59}$$

We obtain the field using Eq. (6.22),

$$E_z(\rho, \phi) = -\frac{k\zeta_0 I_0^C}{4} \sum_{n=0}^{N-1} e^{i2\pi mn/N} H_0^{(1)}\left(k\sqrt{a^2 + \rho^2 - 2a\rho\cos(\phi - 2\pi n/N)}\right) . \tag{6.60}$$

We now apply the PSF-FS (6.51) with

$$g(x) = e^{i2\pi mx/C} H_0^{(1)}\left(k\sqrt{a^2 + \rho^2 - 2a\rho\cos(\phi - 2\pi x/C)}\right) , \tag{6.61}$$

and $C = 2\pi a$ (so that $g(0) = g(C)$). In the integral that arises, set $\phi' = 2\pi x/C = x/a$ to obtain

$$E_z(\rho, \phi)$$
$$= -\frac{k\zeta_0 I_0^C}{4} \frac{N}{2\pi} \sum_{p=-\infty}^{\infty} \int_0^{2\pi} e^{i(m-pN)\phi'} H_0^{(1)}\left(k\sqrt{a^2 + \rho^2 - 2a\rho\cos(\phi - \phi')}\right) d\phi' . \tag{6.62}$$

The integral in Eq. (6.62) was encountered previously in Eq. (6.56). We thus obtain

$$E_z(\rho, \phi) = -\frac{k\zeta_0 I_0^C N}{4} \sum_{p=-\infty}^{\infty} e^{i(m-pN)\phi} \begin{cases} H_{m-pN}^{(1)}(ka) J_{m-pN}(k\rho), & \rho < a \\ \\ J_{m-pN}(ka) H_{m-pN}^{(1)}(k\rho), & \rho > a . \end{cases} \tag{6.63}$$

Note that Eq. (6.63) holds whether Eq. (6.59) is satisfied or not.

Once again, the PSF (in its finite version) has connected the discrete problem with the corresponding continuous one: comparison with Eq. (6.57) reveals that the $p = 0$ term in Eq. (6.63) is the field due to a z-directed, infinite current sheet $K_z^C(\phi')$, where $K_z^C(\phi')$ is given by Eq. (6.53), with m the same as in Eq. (6.58) and with $K_z^C(0) = NI_0^C/(2\pi a)$. The large-$q$ asymptotic approximations of $J_q(x)$ and $H_q^{(1)}(x)$, Eqs. (A.40) and (A.43) of Appendix A, imply that

$$J_q(k\rho_1)H_q^{(1)}(k\rho_2) \sim \frac{-i}{\pi|q|}\left(\frac{\rho_1}{\rho_2}\right)^{|q|} \qquad (q \to \pm\infty) . \qquad (6.64)$$

Thus, when N is large and when $|m|$ is not too close to $N/2$, the remaining terms are all exponentially smaller than the $p = 0$ term and the sum $\sum_{p\neq 0}$ can once again be considered as a remainder; this is especially true when ρ is not too close to a. Similar to Section 6.1.3, the case $|m| = N/2$ is exceptional, because an additional term (the $p = 1$ or the $p = -1$ term) is equal in magnitude to the $p = 0$ term. The additional term must be written together with the $p = 0$ term and must be excluded from the remainder. One must also do this when $|m|$ is very *close* to $N/2$.

Assume for simplicity that $|m|$ is not close to $N/2$. When both $|m|$ and N are large (this is the case relevant to the resonant circular arrays of [9]), observe from Eq. (6.64) that the dominant ($p = 0$) term is exponentially small in $|m|$, so that $E_z(\rho, \phi)$ itself is also exponentially small in $|m|$. Thus, the original representation Eq. (6.60) is not suitable for numerical evaluation, even if Eq. (6.60) involves a *finite* number of terms: summing many large numbers to obtain a small one is always dangerous from the numerical point of view (compare with the similar discussion of small integrals in Section 5.2.4). By contrast, Eq. (6.63) is excellent for numerical evaluation (except when $\rho \cong a$): the remainder is much smaller than the dominant term, and the sum representing the remainder converges very rapidly.

A 3-D plot of the field is given in Fig. 6.4. Eq. (6.63) with a sufficiently large number of terms has been used for the calculation, and the infinities have been truncated. Fig. 6.5 shows the field at $\phi = 0$, calculated in the same manner, but with a logarithmic scale on the vertical axis. Due to roundoff, we were not able to obtain trustworthy results for small $k\rho$, where the field values are extremely small. As a result, only values with $k\rho \geq 12$ are shown. In Fig. 6.5, both the interior and the exterior field initially appear to decrease linearly, indicating a very rapid local initial decrease. The interior field's decrease is much more rapid and appears linear over a wider range than the decrease of the exterior field; for the exterior field, the rate of decrease slows down rapidly around $k\rho = 40$. Ref. [8] discusses a number of additional properties stemming from Eq. (6.63).

6.3   PROBLEMS

6.1.   Starting from Eq. (6.4), derive Eq. (6.1). (The desired derivation is simple, so Eq. (6.1) should not be viewed as being much more general than Eq. (6.1).)

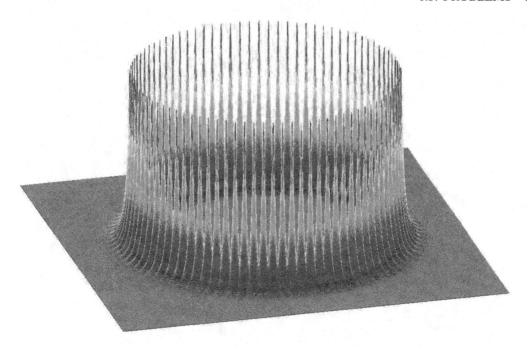

Figure 6.4: Magnitude $|E_z(\rho, \phi)/(kI_0^C)|$ of an electric field due to a cylindrical array of traveling-wave currents; $N = 90$, $m = 40$, and $ka = 27$. *(Figure from Fikioris and Matos [8].)*

6.2. Starting from Eq. (6.1), show Eqs. (6.5) and (6.6).

6.3. Apply Eq. (6.1) for an odd function $f(x)$ ($f(x) = -f(-x)$, $x \in \mathbb{R}$) in order to obtain a formula similar to Eq. (6.5) but involving the sum $\sum\limits_{n=1}^{\infty} f(n) \sin n\theta$.

6.4. Use the PSF and the fact [6] that the Fourier cosine transform of $f(x) = \exp\left(-x^2\right)$ is $\bar{f}_C(y) = \sqrt{\pi} \exp\left(-y^2/4\right)/2$ to show that

$$\sum_{n=-\infty}^{\infty} e^{-xn^2} = \sqrt{\frac{\pi}{x}} \sum_{n=-\infty}^{\infty} e^{-\pi^2 n^2/x}, \quad x > 0. \tag{6.65}$$

Discuss why the two sums have complementary convergence properties. (See [2] for a discussion of Eq. (6.65) within the context of theta functions.)

Figure 6.5: Magnitude $|E_z(\rho,0)/(kI_0^C)|$ of an electric field due to a cylindrical array of traveling-wave currents; $N = 90$, $m = 40$, $ka = 27$, and $\phi = 0$. The scale on the vertical axis is logarithmic. *(Figure from Fikioris and Matos [8].)*

6.5.   PSF in notation of signal processing:

Change the notation in order to rewrite Eq. (6.1) as

$$\sum_{n=-\infty}^{\infty} x(nT_s)e^{-j2\pi fT_s n} = \frac{1}{T_s}\sum_{n=-\infty}^{\infty} X\left(f + \frac{n}{T_s}\right), \quad f \in \mathbb{R}, \tag{6.66}$$

where $T_s > 0$ and

$$X(f) = \int_{-\infty}^{\infty} x(t)\,e^{-j2\pi ft}\,dt \tag{6.67}$$

is the Fourier transform of the signal $x(t)$; note that (6.67) differs from the definition of the Fourier transform normally used in this book. The special case of Eq. (6.66) with $f = 0$ is precisely what is called Poisson's sum formula in [5].

6.6.   Sampling theorem as corollary of PSF:

As stated in [5] and for the special case of sampling at the Nyquist rate, the sampling theorem is: Let the signal $x(t)$ be bandlimited with bandwidth W; i.e., let $X(f) = 0$ for $f \geq W$. Let $x(t)$ be sampled at multiples of the basic sampling interval $T_s = 1/(2W)$, to yield the sequence $x(nT_s)$, $n \in \mathbb{Z}$. Then it is possible to reconstruct the original signal from the sampled values by the reconstruction formula

$$x(t) = \sum_{n=-\infty}^{\infty} x(nT_s) \frac{\sin\left[\pi\left(t/T_s - n\right)\right]}{\pi\left(t/T_s - n\right)} . \tag{6.68}$$

Derive the sampling theorem from the PSF Eq. (6.66) using the following steps:

(i) Show from Eq. (6.66) that under the conditions stated in the sampling theorem we have

$$X(f) = \begin{cases} T_S \displaystyle\sum_{n=-\infty}^{\infty} x(nT_s) e^{-j2\pi f T_s n}, & -1/(2T_s) < f < 1/(2T_s), \\ \\ 0, & |f| \geq 1/(2T_s). \end{cases} \tag{6.69}$$

(ii) Take the inverse Fourier transform of Eq. (6.69) and thus arrive at Eq. (6.68).

6.7.   In this problem, we verify the PSF Eq. (6.4) for the trivial sum $1 = \displaystyle\sum_{n=-\infty}^{\infty} f(n)$ where

$$f(n) = \begin{cases} 1, & n = 0, \\ 0, & n = \pm 1, \pm 2, \dots \end{cases}, \text{ for two cases of interpolating functions } f(x):$$

(i) For $f(x) = \left(\frac{\sin \pi x}{\pi x}\right)^2$, show that the transformed sum (i.e., the sum on the right-hand side of Eq. (6.4)) is the same as the trivial sum we started out with.

(ii) For the interpolating function $f(x) = \begin{cases} 1, & -a < x < a \\ 0, & |x| > a \end{cases}$, where $0 < a < 1$,

show that the transformed sum is $2a + \dfrac{1}{\pi}\displaystyle\sum_{m \neq 0} \frac{\sin(2\pi am)}{m}$, where the sum runs over all nonzero integers.

(iii) Directly sum the quantity in (ii) and verify that it equals the correct answer 1. *Hint:* Use Eq. (3.29).

6.8.   With the aid of a contour integral, derive Eq. (6.13).

6.9.   Verify that the sequence in Eq. (6.16) is an asymptotic sequence as $\tau \to 0$.

6.10.   Verify that, with the ordering of Eq. (6.18), the sequence in Eq. (6.17) is an asymptotic sequence as $\tau \to 0$.

6.11. Verify numerically that, for small ϕ and τ, Eq. (6.20) is better than Eq. (6.19). (The exact value is easily calculated using a sufficient number of terms in Eq. (6.14); it is also possible to use Eq. (6.7).)

6.12. Starting from Eq. (6.33) and subject to Eq. (6.27), show Eq. (6.35). Discuss why Eq. (6.35) *does not* hold when $\beta d = \pi$. What modification is then necessary?

6.13. Give a nonrigorous derivation of Eq. (6.40)–(6.42) via the PSF for semi-infinite sums. *Hint:* Apply Eq. (6.6) for $f(x) = g(xC/N)$ and again for $f(x) = g(xC/N + C)$.

6.14. Verify the PSF-FS for (i) $g(n) = 1$ and (ii) $g(n) = e^n$.

REFERENCES

[1] P. M. Morse and P. Feshbach, *Methods of Theoretical Physics, Part I*. New York: McGraw-Hill, 1953. 106

[2] T. M. Apostol, *Mathematical Analysis: A Modern Approach to Advanced Calculus, 2nd Ed.* New York: Pearson, 1974, §11.22. 106, 123

[3] F. W. J. Olver, D. W. Lozier, R. F. Boisvert, and C. W. Clark, *Digital Library of Mathematical Functions*, National Institute of Standards and Technology from `http://dlmf.nist.gov/` , §1.8.14, §1.8.15, §10.21.3. 106, 107, 121

[4] I. Stakgold and M. Holst, *Green's functions and Boundary Value Problems, 3rd Ed.* New York: John Wiley, 2011. DOI: 10.1002/9780470906538. 106

[5] J. G. Proakis and M. Salehi, *Communication System Engineering, 2nd Ed.* Upper Saddle River, NJ: Prentice-Hall, 2002, p. 46, p. 63. 107, 124, 125

[6] F. Oberhettinger, *Tables of Fourier Transforms and Fourier Transforms of Distributions*. Berlin: Springer-Verlag, 1990, Entries 1.3.17, 1.5.83, 1.5.88, 1.7.1, 1.14.16, 1.14.18. DOI: 10.1007/978-3-642-74349-8. 107, 108, 112, 115, 123

[7] N. N. Lebedev, I. P. Skalskaya, and Y. S. Uflyand, *Worked Problems in Applied Mathemics*. New York: Dover, 1965, Problems 190, 192, 271. 108

[8] G. Fikioris and K. Matos, "Near fields of resonant circular arrays of cylindrical dipoles," *IEEE Antennas and Propagation Magazine*, vol. 50, no. 1, pp. 97–107, Feb. 2008. DOI: 10.1109/MAP.2008.4494508. 110, 111, 113, 114, 119, 120, 122, 123, 124

[9] R. W. P. King, G. Fikioris, and R. B. Mack, *Cylindrical Antennas and Arrays*. Cambridge, UK, Cambridge University Press, 2002, Chapters 11 and 12. DOI: 10.1017/CBO9780511541100. 110, 113, 122

[10] C. A. Balanis, *Advanced Engineering Electromagnetics*. New York, Wiley, 1989, §5.3.4. 112

[11] I. Chremmos and G. Fikioris, "Traveling waves in two parallel infinite linear point-scatterer arrays," *J. Opt. Soc. Am. A.*, vol. 28, no. 6, pp. 962–969, 2011. DOI: 10.1364/JOSAA.28.000962. 114, 115

[12] I. Psarros and G. Fikioris, "Two-term theory for infinite linear array and application to study of resonances," *J. of Electromagnetic Waves and Applications*, vol. 20, no. 5, pp. 623–645, April 2006. DOI: 10.1163/156939306776137809. 114, 115

[13] L. B. Felsen and E. G. Ribas, "Ray theory for scattering by two-dimensional quasiperiodic plane finite arrays," *IEEE Trans. Antennas Propagat.*, vol. 44, no. 3, pp. 375–382, March 1996. DOI: 10.1109/8.486307. 116

[14] L. B. Felsen and L. Carin, "Diffraction theory of frequency- and time-domain scattering by weakly aperiodic truncated thin-wire gratings," *J. Opt. Soc. Amer. A*, vol. 11, no. 4, pp. 1291–1306, April 1994. DOI: 10.1364/JOSAA.11.001291. 116

[15] G. Fikioris, S. D. Zaharopoulos, and P. D. Apostolidis, "Field patterns of resonant closed-loop arrays: Further analysis," *IEEE Trans. Antennas Propagat.*, vol. 53, no. 12, pp. 3906–3914, Dec. 2005. DOI: 10.1109/TAP.2005.859760. 116

[16] P. J. Davis and P. Rabinowitz, *Methods of Numerical Integration, 2nd Ed.* Orlando, FL: Academic Press, 1984, §2.9, §4.6.5. 116, 119

[17] J. W. Brown and R. V. Churchill, *Fourier Series and Boundary Valued Problems, 5th Ed.* New York: McGraw-hill International Ed., 1993, pp. 73–89. 117

[18] Ö. A. Çivi, P. H. Pathak, and H.–T. Chou, "On the Poisson sum formula for the analysis of wave radiation and scattering from large finite arrays," *IEEE Trans. Antennas Propagat.*, vol. 47, no. 5, pp. 958–959, May 1999. DOI: 10.1109/8.774163. 118

[19] R. Kress, *Numerical Analysis*. New York: Springer, 1998, §9.4. DOI: 10.1007/978-1-4612-0599-9. 119

[20] B. Z. Katsenelenbaum, "Nonapproximable diagrams and nonradiating currents," *Journal of Communications Technology and Electronics*, vol. 38, no. 11, pp. 112–118, 1993 (originally published in *Radiotekhnika i elektronika*, no. 6, pp. 998-1005, 1993). 121

[21] B. Z. Katsenelenbaum, *Electromagnetic fields: Restrictions and approximations*. Weinheim, Germany:Wiley, VCH, 2003. DOI: 10.1002/9783527602568.

[22] D. Margetis, G. Fikioris, J. M. Myers, and T. T. Wu, "Highly directive current distributions: General theory," *Physical Review E*, vol. 58, no. 2, pp. 2531–2547, Aug. 1998. DOI: 10.1103/PhysRevE.58.2531.

[23] A. J. Devaney and E. Wolf, "Radiating and nonradiating classical current distributions and the fields they generate," *Physical Review D,* vol. 8, no. 4, pp. 1044–1047, 15 Aug. 1973. DOI: 10.1103/PhysRevD.8.1044. 121

CHAPTER 7

Mellin-Transform Method for Asymptotic Evaluation of Integrals

In a previous Synthesis Lecture [1], we described the Mellin-transform method for the *exact* evaluation of certain integrals. The purpose of the present chapter is to briefly summarize this method (readers completely unfamiliar with it can consult [1]) and to indicate the necessary modifications that enable us to obtain asymptotic expansions. The said modifications are minor and the method is very powerful: it can handle a wide class of integrals in the form

$$f(x) = \int_a^b g(xy)h(y) \; dy \,, \quad 0 \le a < b \le \infty \,, \quad x > 0 \,, \tag{7.1}$$

often giving full asymptotic expansions both for $x \to 0^+$ and for $x \to +\infty$. While the required calculations may be laborious, the method is frequently straightforward to apply.

7.1 SUMMARY OF MELLIN-TRANSFORM METHOD

The Mellin transform of $f(x)$ is defined as

$$\text{M.T.} \{f(x); z\} = \widetilde{f}(z) = \int_0^\infty x^{z-1} f(x) \; dx, \quad z \in \text{SOA} \left\{ \widetilde{f}(z) \right\} \,. \tag{7.2}$$

In Eq. (7.2), z must be restricted to those complex values for which the integral converges (if it converges at all). As one can easily understand from Appendix B (in particular, from the examples in Section B.3), convergence occurs at $x = 0$ only if $\Re z$ is larger than a certain value, and at $x = +\infty$ only if $\Re z$ is smaller than a certain value. Thus, if the Mellin transform of $f(x)$ (as defined in Eq. (7.2)) exists at all, it exists in a vertical strip in the complex-z plane. This is called the "strip of analyticity" of $\widetilde{f}(z)$ and is denoted here by SOA $\left\{ \widetilde{f}(z) \right\}$ or, more simply, SOA.

The following properties of the Mellin transform are elementary,

$$\text{M.T.} \{f(\alpha x); z\} = \alpha^{-z} \widetilde{f}(z), \quad \alpha > 0, \tag{7.3}$$

$$\text{M.T.} \{x^{\alpha} f(x); z\} = \widetilde{f}(z + \alpha), \tag{7.4}$$

$$\text{M.T.} \{f(x^{\alpha}); z\} = \frac{1}{|\alpha|} \widetilde{f}\left(\frac{z}{\alpha}\right), \quad \alpha \in \mathbb{R} - \{0\}. \tag{7.5}$$

These can be used together with published tables to find the Mellin transform of a given function. By far the most extensive table is Section 8.4 in Volume 3 of the work by Prudnikov, Brychkov, and Marichev [2]. Mellin transforms can also be found using symbolic programs and online in the Wolfram Functions Site [3].

The Mellin transform of "most" functions $f(x)$ can be written as a linear combination of what [1] calls "standard products." These consist of α^{-z} times a product of factors having the form $\Gamma(a_n + A_n z)$ or $[\Gamma(b_n + B_n z)]^{-1}$, where all A_n and B_n are real. For example, the Mellin transform and SOA of the Bessel function $J_{\nu}(x)$ is [1]

$$\text{M.T.} \{J_{\nu}(x); z\} = \frac{1}{2}\left(\frac{1}{2}\right)^{-z} \frac{\Gamma\left(\frac{\nu}{2} + \frac{z}{2}\right)}{\Gamma\left(1 + \frac{\nu}{2} - \frac{z}{2}\right)}, \quad -\Re\nu < \Re z < \frac{3}{2}. \tag{7.6}$$

The aforementioned table in [2] is particularly useful for the Mellin-transform method because it always expresses its Mellin transforms as standard products; as noted in [1], not all tables do so.

The Mellin inversion formula is

$$f(x) = \frac{1}{2\pi i} \int_{\delta - i\infty}^{\delta + i\infty} x^{-z} \widetilde{f}(z) \, dz, \quad \delta \in \text{SOA} \left\{\widetilde{f}(z)\right\}, \tag{7.7}$$

where $\delta \in \mathbb{R}$ and the vertical integration path lies within the SOA, as shown in Fig. 7.1. When a Mellin transform has the form of a standard product, the integrand of the inversion integral will also be a standard product multiplied by x^{-z}. When applied to Eq. (7.6), for example, Eq. (7.7) gives

$$J_{\nu}(x) = \frac{1}{2} \frac{1}{2\pi i} \int_{\delta - i\infty}^{\delta + i\infty} \left(\frac{x}{2}\right)^{-z} \frac{\Gamma\left(\frac{\nu}{2} + \frac{z}{2}\right)}{\Gamma\left(1 + \frac{\nu}{2} - \frac{z}{2}\right)} \, dz, \quad -\Re\nu < \delta < \frac{3}{2}. \tag{7.8}$$

Convergent integrals like the one in the right-hand side of Eq. (7.8)—with integrands of the aforementioned type, integrated along proper contours in the z-plane—are called *Mellin-Barnes integrals*. Therefore, "most" functions $f(x)$ can be written as Mellin-Barnes integrals, or as linear combinations of Mellin-Barnes integrals.

The *Mellin convolution* of $g(x)$ and $h(x)$ is defined as

$$(g \emptyset h)(x) = \int_0^{\infty} g(xy)h(y) \, dy, \quad x > 0. \tag{7.9}$$

Figure 7.1: Integration path in the inversion formula (7.7).

Corresponding to the familiar convolution theorems pertaining to the Fourier and Laplace transforms, for the Mellin transform it is true that

$$\text{M.T.}\left\{g(x)\emptyset h(x); z\right\} = \widetilde{g}(z)\widetilde{h}(1-z), \quad z \in \text{SOA}\left\{\widetilde{g}(z)\right\} \cap \text{SOA}\left\{\widetilde{h}(1-z)\right\}. \quad (7.10)$$

In the above equation, z must belong to both SOA $\left\{\widetilde{g}(z)\right\}$ and SOA $\left\{\widetilde{h}(1-z)\right\}$—for Eq. (7.10) to hold, the two strips should ordinarily overlap.

As already mentioned, the Mellin-transform method is useful for certain $f(x)$ that have the form Eq. (7.1) or, in other words, to certain integrals that are Mellin convolutions. The first step of the method is to obtain a complex-integral representation of $f(x)$ by applying the convolution theorem Eq. (7.10) and the inversion formula Eq. (7.7),

$$f(x) = \frac{1}{2\pi i} \int_{\delta-i\infty}^{\delta+i\infty} x^{-z}\widetilde{g}(z)\widetilde{h}(1-z)\,dz, \quad \delta \in \text{SOA}\left\{\widetilde{g}(z)\right\} \cap \text{SOA}\left\{\widetilde{h}(1-z)\right\}. \quad (7.11)$$

The integral will, hopefully, be a Mellin-Barnes integral. In Eq. (7.11), the vertical contour belongs both to SOA $\left\{\widetilde{g}(z)\right\}$ and to SOA $\left\{\widetilde{h}(1-z)\right\}$, which must overlap. Equation (7.11) is the heart of the Mellin-transform method and, having obtained it, one can proceed in many ways.

First, Eq. (7.11) can often yield a series expansion. Use the familiar from Chapter 3 properties of the gamma function to analytically continue the integrand beyond the SOA. Then, determine the singularities of the integrand to the left (or right) of the contour which—as long as

Eq. (7.11) is a Mellin-Barnes integral—will be poles (but not necessarily *simple* poles), located at $z = z_0, z_1, \ldots$. Then, close the contour at left (or right, respectively) and apply the residue theorem to obtain

$$f(x) = \pm \sum_{n=0}^{\infty} \operatorname*{res}_{z=z_n} \left[\widetilde{g}(z)\widetilde{h}(1-z)x^{-z} \right] , \tag{7.12}$$

where the upper (lower) sign corresponds to closing the contour at left (right). Sufficient conditions enabling one to close the contour, and sufficient criteria for the convergence of the series in Eq. (7.12), are given in [1]. Often, the series can be identified with a generalized hypergeometric function $_pF_q$; this is a class of special functions discussed briefly in Section A.6 of Appendix A.

Alternatively, one can proceed directly from the Mellin-Barnes integral representation Eq. (7.11) to obtain an expression for $f(x)$ as an even more general class of functions, the so-called Meijer-G function [1]. The $_pF_q$ or G-function can then simplify to one of the more usual special functions of mathematical physics; as discussed in [1], useful tables for this purpose are Chapter 7 of [2], Section 8.4.52 of [2], and the online table in the Wolfram Functions site [3].

Thus, one way to generate an asymptotic approximation of $f(x)$ is to first use the Mellin-transform method—as described in [1]—to find a closed-form evaluation, and to then look up the asymptotic approximation of the obtained special function. This may work for non-simple special functions—and, in real problems, may even give an asymptotic expansion in terms of a variable other than x—because there is extensive literature on the asymptotics of $_pF_q$ and G-functions: we can, for example, consult [3, 4, 5, 6], or a growing number of research papers.

Rather than producing a closed-form expression, our interest in what follows is to demonstrate that, with the lower sign, the series in Eq. (7.12) may itself be a (divergent) large-x asymptotic expansion; for this purpose, Eq. (7.12) is better viewed as resulting not from the procedure of closing the contour, but from shifting the contour indefinitely to the right, and picking up residues from the z_n. For simplicity, we assume that the z_n in Eq. (7.12) are simple or double poles to the right of the contour. We begin by giving two formulas useful for calculating residues.

7.2   LEMMAS FOR RESIDUE CALCULATIONS

The residue of a standard product at a simple pole can be found using the following lemma [1]:

Lemma 7.1   *(useful for simple poles)*
   When $\alpha \neq 0$ and $x > 0$, the singularities of $x^{-z}\Gamma(\alpha z + \beta)$ are simple poles located at

$$p_n = -\frac{\beta + n}{\alpha}, \quad n = 0, 1, 2, \ldots , \tag{7.13}$$

with corresponding residues

$$\operatorname*{res}_{z=p_n} \left[x^{-z}\Gamma(\alpha z + \beta) \right] = \frac{1}{\alpha}\frac{(-1)^n}{n!}x^{-p_n}, \quad n = 0, 1, 2, \ldots, \quad \alpha \neq 0 . \tag{7.14}$$

The proof of Lemma 7.1 is very simple (Problem 7.1). For the case of double poles, the following lemma can often (but not always) be used.

Lemma 7.2   *(often useful for double poles) [1]*
 If $x > 0$, $n = 0, 1, 2, \ldots$, and $g(z)$ is analytic and nonzero at $z = -n$, then $[\Gamma(z)]^2 g(z) x^{-z}$ has a double pole at $z = -n$ and the residue there is

$$\operatorname*{res}_{z=-n} \left\{ x^{-z} \left[\Gamma(z)\right]^2 g(z) \right\} = \frac{x^n}{(n!)^2} \left[-g(-n)\ln x + 2\psi(n+1)g(-n) + g'(-n)\right] ,$$

(7.15)

$$n = 0, 1, \ldots ,$$

where ψ is the psi function defined in Eq. (3.12).

Lemma 7.2 is shown in Appendix B of [1] (the proof will not be repeated here, but we note that it is good practice with the gamma-related functions of Chapter 3). Many expressions arising when applying the Mellin-transform method can be written in a form appropriate for the application of Lemma 7.2. For example, to find the residues of $\Gamma(z - m)\Gamma(z)h(z)x^{-z}$ ($h(z)$ analytic and nonzero at $z = -m$, $m = 0, 1, 2, \ldots$), use Lemma 7.2 after substituting $\Gamma(z - m)$ by $\Gamma(z)/(z - m)_m$, where $(z)_m$ is Pochhammer's symbol, see Eq. (3.17).

Note the appearance of $\ln x$ in Eq. (7.15). Lemmas 7.1 and 7.2 already indicate why Eq. (7.12) leads to an ascending series and, in the case of simple poles, a formal power series. This will be better understood from the examples that follow.

7.3   SIMPLE EXAMPLE

Consider the integral

$$f(x) = \int_0^\infty \frac{J_0(yx)}{\sqrt{1 + y^2}}\, dy ,$$

(7.16)

which is the Mellin convolution

$$f(x) = g(x) \emptyset h(x) = J_0(x) \emptyset \frac{1}{\sqrt{1 + x^2}} .$$

(7.17)

The Mellin transform of $g(x)$ is a special case of Eq. (7.6),

$$\tilde{g}(z) = \frac{1}{2}\left(\frac{1}{2}\right)^{-z} \frac{\Gamma\left(\frac{z}{2}\right)}{\Gamma\left(1 - \frac{z}{2}\right)} , \quad 0 < \Re z < \frac{3}{2} ,$$

(7.18)

and that of $h(x)$ can be found from Entry 8.4.2.5 of [2] and Eq. (7.5). The quantity $\tilde{h}(1-z)$ turns out to be

$$\tilde{h}(1-z) = \frac{1}{2\sqrt{\pi}}\Gamma\left(\frac{1-z}{2}\right)\Gamma\left(\frac{z}{2}\right), \quad 0 < \Re z < 1. \tag{7.19}$$

Thus, the desired Mellin-Barnes integral representation Eq. (7.11) is

$$f(x) = \frac{1}{2\pi i}\int_{\delta-i\infty}^{\delta+i\infty}\left(\frac{x}{2}\right)^{-z}\frac{1}{4\sqrt{\pi}}\frac{\left[\Gamma\left(\frac{z}{2}\right)\right]^2\Gamma\left(\frac{1-z}{2}\right)}{\Gamma\left(1-\frac{z}{2}\right)}\,dz, \quad 0 < \delta < 1. \tag{7.20}$$

Our interest is in the poles to the right of the contour. By Eq. (7.13), these are simple poles at $z = 1, 3, 5, \ldots$, contributed by $\Gamma\left((1-z)/2\right)$ and denoted by an "s" in Fig. 7.2. The other two gamma functions give poles to the left of the contour (denoted by "d" in Fig. 7.2) and zeros to the right of the contour—these zeros, importantly, do not cancel the aforementioned poles at $z = 1, 3, 5, \ldots$.

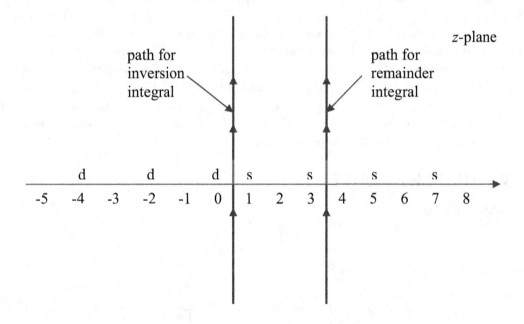

Figure 7.2: Simple (s) and double (p) poles of integrand in (7.20), integration path for inversion integral in (7.20), and path for remainder integral in (7.21).

Shifting the contour two places to the right as shown in Fig. 7.2 (we further discuss this step in the next section), we pick up contributions from the poles at $z = 1$ and $z = 3$. By Eq. (7.14), the corresponding residues are $-1/x$ and $-1/\left(2x^3\right)$. Thus,

$$f(x) = \frac{1}{x} + \frac{1}{2x^3} + \text{remainder integral} . \tag{7.21}$$

Continuation of this process (and some algebra) yields

$$f(x) = \frac{1}{2\pi^{3/2}} \sum_{n=0}^{N} \frac{\left[\Gamma\left(n + \frac{1}{2}\right)\right]^3}{n!} \left(\frac{2}{x}\right)^{2n+1} + \text{remainder integral} . \tag{7.22}$$

If we let $N \to \infty$, we obtain a series which does not converge for any (large) value of x. However, we can write

$$f(x) \sim \frac{1}{2\pi^{3/2}} \sum_{n=0}^{\infty} \frac{\left[\Gamma\left(n + \frac{1}{2}\right)\right]^3}{n!} \left(\frac{2}{x}\right)^{2n+1} , \quad \text{as} \quad x \to \infty . \tag{7.23}$$

This is the complete asymptotic expansion of $f(x)$ for large x. The reader may wish to verify Eq. (7.23) by comparing it with results in the literature (Problem 7.3).

7.4   ON THE CONVERGENCE OF MELLIN-BARNES INTEGRALS

In this section, we give, without proof, some results from [7] that verify the convergence of the remainder integrals in Eqs. (7.21) and (7.22). (Our treatment here is slightly different from that of Chapter 3 of [1]: the latter results are expressed in terms of the G-function, which we do not use here.)

Consider the Mellin-Barnes integral

$$\frac{1}{2\pi i} \int_{\delta-i\infty}^{\delta+i\infty} x^{-z} \frac{\prod_{j=1}^{m} \Gamma\left(a_j + A_j z\right) \prod_{j=1}^{n} \Gamma(b_j - B_j z)}{\prod_{j=1}^{p} \Gamma(c_j + C_j z) \prod_{j=1}^{q} \Gamma(d_j - D_j z)} \, dz , \tag{7.24}$$

where

1. $\delta \in \mathbb{R}$ and $x > 0$;

2. all z-coefficients are strictly positive, i.e., $A_j > 0$, $B_j > 0$, $C_j > 0$, and $D_j > 0$;

3. a_j, b_j, c_j, $d_j \in \mathbb{R}$;

4. the path of integration does not pass through any poles of the integrand;

5. an empty product is to be interpreted (in the usual manner) as 1.

Assumptions 1 and 2 imply that all poles are on the real axis, which is the most usual case. We identify some cases—derived in [7] on the basis of Stirling's formula—where shifting the contour is allowed and some cases where this is not allowed. We do not cover all cases.

Define the quantities

$$\alpha = \sum_{j=1}^{m} A_j + \sum_{j=1}^{n} B_j - \sum_{j=1}^{p} C_j - \sum_{j=1}^{q} D_j \tag{7.25}$$

and

$$\beta = \sum_{j=1}^{m} A_j - \sum_{j=1}^{n} B_j - \sum_{j=1}^{p} C_j + \sum_{j=1}^{q} D_j . \tag{7.26}$$

- The Mellin-Barnes integrals Eq. (7.24) is of the *first type* if $\alpha > 0$. Mellin-Barnes integrals of the first type converge for all δ (i.e., for all vertical contours).

- The Mellin-Barnes integral Eq. (7.24) is of the *second type* if $\alpha = 0$ and $\beta \neq 0$.

  – If $\beta > 0$, Eq. (7.24) converges only if δ is smaller than a certain value (provided in [7]). Thus, we cannot shift the contour to the right indefinitely.

  – If $\beta < 0$, Eq. (7.24) converges only if δ is larger than a certain value (provided in [7]). Thus, we cannot shift the contour to the left indefinitely.

It is easily checked that Eq. (7.20) is a Mellin-Barnes integral of the first type, so that the shifting of the contour is justified. We stress that checking the convergence is an important step: Problem 7.5 shows that blind application of the "method" can give erroneous results.

7.5   APPLICATION TO HIGHLY DIRECTIVE CURRENT DISTRIBUTIONS

Two-dimensional, continuous current distributions placed on a circular cylinder of electric radius ka (where k is the free-space wavenumber) were encountered in Chapter 6. Ref. [8] examines current distributions of this sort that possess two properties: (i) they are optimum in the sense that they maximize the directivity D for fixed $C = N/T$, where N is the integral of the square of the current magnitude and T is the total radiated power, and (ii) they are highly directive in the sense that the associated D is larger than a certain threshold D_0. The quantity D_0 is the directivity in the familiar *reference case*, in which currents of uniform magnitude produce the maximum directivity by constructive interference. In [8], interest is mainly in large ka and in both moderate and large values of C. Moderate values correspond to the practical case of directivities slightly larger than D_0, while the limit $C \to \infty$ corresponds to Oseen's "Einstein needle radiation" [8, 9].

In this section, we confine our interest to $T_0(ka)$, the total radiated power in the reference case. This is given by the integral (Eq. (3.5) of [8])

$$T_0(ka) = \frac{2}{\pi} \int_0^{\pi/2} [J_0(2ka\sin\phi)]^2 d\phi = \frac{2}{\pi} \int_0^1 \frac{[J_0(2kay)]^2}{\sqrt{1-y^2}} \, dy \,, \qquad (7.27)$$

which can readily be calculated numerically. Also, straightforward application of the Mellin-transform method—as described in [1]—yields a closed-form evaluation of $T_0(ka)$ in terms of the generalized hypergeometric function $_2F_3$, see Eq. (3.33) of [8] or the more general Eq. (5.7) of [1]. Here, we use Mellin transforms to find the leading term of the asymptotic expansion of $T_0(ka)$. We then briefly describe how to supplement this term in order to obtain a simple and informative asymptotic approximation for the case $ka \gg 1$.

We can (Problem 7.6) easily obtain the Mellin-Barnes integral representation

$$T_0(ka) = \frac{1}{2\pi i} \int_{\delta - i\infty}^{\delta + i\infty} (2ka)^{-z} \frac{1}{2\pi} \frac{\left[\Gamma\left(\frac{1-z}{2}\right)\right]^2 \Gamma\left(\frac{z}{2}\right)}{\left[\Gamma\left(1 - \frac{z}{2}\right)\right]^3} \, dz \,, \quad 0 < \delta < 1 \,. \qquad (7.28)$$

The singularities to the right of the contour are *double* poles at $z = 1, 3, 5, \ldots$. Using Lemma 7.2 and retaining only the contribution from $z = 1$, we find (Problem 7.7) the asymptotic approximation

$$T_0(ka) \sim \frac{\ln(32ka) + \gamma}{\pi^2 ka} \,, \qquad (7.29)$$

where γ is Euler's constant, discussed in Chapter 3.

Because the integral in Eq. (7.28) is a Mellin-Barnes integral of the second type, we cannot obtain the full asymptotic expansion by shifting the contour indefinitely to the right. Using a rather complicated procedure, [8] shows that $T_0(ka)$ possesses a compound asymptotic expansion, described by Eqs. (3.11), (3.15), and (3.28) of [8]. Describing this procedure is beyond the scope of this book. Instead, we supplement Eq. (7.29) with the "steepest descent contribution" that can be found in the systematic treatment of Mellin-Barnes integrals in the work [10] of Sasiela: we immediately obtain the improved asymptotic approximation

$$T_0(ka) \sim \frac{\ln(32ka) + \gamma}{\pi^2 ka} - \frac{1}{(2\pi ka)^{3/2}} \cos\left(4ka + \frac{\pi}{4}\right) \,, \qquad (7.30)$$

which agrees with the first two terms in Eq. (3.30) of [8]. With Eq. (7.30), we clearly understand the behavior of T_0 as ka is varied. As demonstrated by Figs. 7.3 and 7.4, Eq. (7.29) does not adequately capture the variations of T_0, while Eq. (7.30) achieves excellent accuracy. The relative error is $\leq 0.3\%$ for $ka \geq 5$ and $\leq 10\%$ for $ka \geq 0.5$. Note that the values of ka in Fig. 7.4 are quite small; since Eq. (7.30) is a large-ka approximation, the agreement exhibited in that figure is rather surprising.

Figure 7.3: $T_0(ka)$ as calculated numerically by (7.27) (solid line) *and* by the approximate formula (7.30) (which includes the cosine): at the scale of the figure, the two curves coincide. The dashed line shows corresponding results obtained via (7.29) (which does not include the cosine).

7.6   FURTHER READING

The discussions of this chapter are meant to serve as a brief introduction to a method that, as already mentioned, is very powerful and often underutilized. For example, when discussing the method, Ablowitz and Fokas [11] state, "This method, although often quick and easy to apply, is not widely known." For descriptions of the method in more detail or with more rigor, see [6], [10], [12]–[15].

The integral Eq. (7.27) and certain generalizations are relevant to other antenna problems [1] and to applications in crystallography and diffraction theory. The said integrals have been the subject of much discussion in both the mathematical and the antenna/electromagnetics literature: see the relevant references cited in Section 5.2 of [1].

7.7   PROBLEMS

7.1.   Prove Lemma 7.1.

Figure 7.4: $T_0(ka)$ computed numerically by (7.27) (solid line), by approximate formula (7.29) (which does not include the cosine) (dot-dashed line), and by approximate formula (7.30) (which includes the cosine) (dashed-line). It is seen that, surprisingly, the large-ka formula (7.30) is suitable even for small values of ka.

7.2. Using Lemma 7.2, show that (in addition to a simple pole at $z = 1$), $x^{-z}\Gamma(z)\Gamma(z - 1)$ has double poles at $z = 0, -1, -2, \ldots$, with residues

$$\operatorname*{res}_{z=-n}\left\{x^{-z}\Gamma(z)\Gamma(z - 1)\right\} = \frac{x^n}{n!(n + 1)!}\left[\ln x - 2\psi(n + 1) - \frac{1}{n + 1}\right], \quad n = 0, 1, \ldots$$

7.3. Let $f(x)$ be the integral defined in Eq. (7.16).

(i) By table lookup, or by using a symbolic routine, verify that $f(x) = I_0(x/2)K_0(x/2)$.

(ii) Check that Eq. (7.23) is consistent with the asymptotic expansion of $I_0(x)K_0(x)$ given in [6].

7.4. Elsewhere in this book, we showed that $J_0(x)$ has a compound asymptotic expansion consisting of the equality Eq. (2.10), viz.,

$$J_0(x) = \frac{2}{\pi} f_1(x) \sin x + \frac{2}{\pi} f_2(x) \cos x, \quad x > 0,$$

together with asymptotic power series for the two functions $f_1(x)$ and $f_2(x)$, with the functions defined by the integrals in Eq. (2.11). Proceeding from Eq. (2.11) and using the Mellin-transform method, find the two asymptotic power series. Compare your results to Section 4.2.3.

7.5. Consider the integral

$$f(x) = \int\limits_0^1 \frac{J_0(yx)}{\sqrt{1-y^2}} \, dy, \quad x > 0.$$

(i) Show that the Mellin transform and corresponding SOA are

$$\tilde{f}(z) = \frac{\sqrt{\pi}}{4} \left(\frac{1}{2}\right)^{-z} \frac{\Gamma\left(\frac{z}{2}\right) \Gamma\left(\frac{1-z}{2}\right)}{\left[\Gamma\left(1-\frac{z}{2}\right)\right]^2}, \quad 0 < \Re z < 1.$$

(ii) Explain why blind application of our method (i.e., shifting the contour indefinitely to the right) would result in an asymptotic power series for $f(x)$ as $x \to +\infty$.

(iii) By table lookup, verify the exact answer $f(x) = \frac{\pi}{2}[J_0(x)]^2$. With the aid of Appendix A, explain why $f(x)$ admits a compound asymptotic expansion involving sines and cosines rather than an asymptotic power series. It follows that blind application of our method would yield erroneous results.

7.6. Show Eq. (7.28).

7.7. Use Lemma 7.2 to verify Eq. (7.29).

REFERENCES

[1] G. Fikioris, *Mellin-transform method for integral evaluation: Introduction and applications to electromagnetics.* (Synthesis Lectures on Computational Electromagnetics #13). Morgan and Claypool Publishers, 2007. DOI: 10.2200/S00076ED1V01Y200612CEM013. 129, 130, 132, 133, 135, 137, 138

[2] A. P. Prudnikov, Yu. A. Brychkov, and O. I. Marichev, *Integrals and Series: More Special Functions*, vol. 3. London, UK: Taylor and Francis, 2002. (Reprint of 1990 Ed.) 130, 132, 134

[3] The Wolfram Functions Site from `http://functions.wolfram.com/` . 130, 132

[4] Y. L. Luke, *The Special Functions and Their Approximations,* vols. I and II. New York: Academic Press, 1969. 132

[5] Y. L. Luke, *Mathematical Functions and their Approximations.* New York: Academic Press, 1975. 132

[6] F. W. J. Olver, D. W. Lozier, R. F. Boisvert, and C. W. Clark, *Digital Library of Mathematical Functions,* National Institute of Standards and Technology from `http://dlmf.nist.gov/` , Chapters 2 and 16. 132, 138, 139

[7] Staff of the Bateman Manuscript Project (A. Erdélyi (Editor), W. Magnus, F. Oberhettinger, and F. G. Tricomi (Research Associates)), *Higher Transcendental Functions. Vol. I.* New York: McGraw-Hill, 1953. (Reprinted, Malabar, FL: Robert Krieger Publishing Co., 1981), §1.19. (Note that the z^s in eqn. (1) of §1.19 must be replaced by z^{-s}.) 135, 136

[8] D. Margetis and G. Fikioris, "Two-dimensional, highly directive currents on large circular loops," *Journal of Mathematical Physics,* vol. 41, no. 9, pp. 6130–6172, September 2000. DOI: 10.1063/1.1288245. 136, 137

[9] D. Margetis, G. Fikioris, J. M. Myers, and T. T. Wu, "Highly directive current distributions: General theory," *Physical Review E,* vol. 58, no. 2, pp. 2531–2547, August 1998. DOI: 10.1103/PhysRevE.58.2531. 136

[10] R. J. Sasiela, *Electromagnetic wave propagation in turbulence: evaluation and application of Mellin transforms, 2nd Ed.* Bellington, Washington: SPIE Press, 2007. DOI: 10.1117/3.741372. 137, 138

[11] M. J. Ablowitz and A. S. Fokas, *Complex Variables: Introduction and Applications, 2nd Ed.* Cambridge, UK: Cambridge University Press, 2003. DOI: 10.1017/CBO9780511791246. 138

[12] R. B. Paris and D. Kaminski, *Asymptotics and Mellin-Barnes Integrals.* Cambridge, UK: Cambridge University Press, 2001. DOI: 10.1017/CBO9780511546662. 138

[13] N. Bleistein and R. A. Handelsman, *Asymptotic Expansions of Integrals.* New York: Dover, 1986.

[14] R. Wong, *Asymptotic Approximations of Integrals.* Philadelphia: SIAM, 2001. DOI: 10.1137/1.9780898719260.

[15] F. W. J. Olver, *Asymptotics and Special Functions.* Natick, MA: A. K. Peters, 1997. 138

CHAPTER 8

More Applications to Wire Antennas

The purpose of this last chapter is to present some additional applications, related to fundamental aspects of computational methods for wire antennas, that combine techniques developed in the previous chapters. An $e^{-i\omega t}$ time dependence is assumed throughout, with $k = \omega/c = 2\pi/\lambda$.

8.1 PROBLEM PERTAINING TO MAGNETIC FRILL GENERATOR

8.1.1 STATEMENT OF PROBLEM

For the case of a linear antenna of length $2h$ and radius a driven by a magnetic frill generator, Pocklington's equation is Eq. (3.54). When $h = \infty$ so that the antenna is infinite in length, Eq. (3.54) becomes

$$\left(\frac{\partial^2}{\partial z^2} + k^2\right) \int_{-\infty}^{\infty} K(z - z')I(z') \, dz' = g(z), \quad z \in \mathbb{R}, \tag{8.1}$$

where

$$g(z) = \frac{ikV}{2\zeta_0 \ln(b/a)} \left[\frac{\exp\left(ik\sqrt{z^2 + a^2}\right)}{\sqrt{z^2 + a^2}} - \frac{\exp\left(ik\sqrt{z^2 + b^2}\right)}{\sqrt{z^2 + b^2}} \right], \tag{8.2}$$

in which b is the outer radius of the frill, V is the equivalent to the frill driving voltage, $\zeta_0 = 376.73$ Ohms, and the approximate kernel $K(z)$ is given in Eq. (3.51). By Eq. (1.2), the integral in (8.1) is—apart from a multiplicative constant—the vector potential $A_z(z)$ and is an even function of z. Solving Eq. (8.1) for the integral, we obtain Hallén's equation,

$$\int_{-\infty}^{\infty} K(z - z')I(z') \, dz' = \frac{1}{k} \int_{0}^{z} g(t) \sin k(z - t) \, dt + C \cos kz, \tag{8.3}$$

where C is a constant. To determine C, we demand that the right-hand side of Eq. (8.3) must represent an outgoing wave when $|z| \to \infty$. This leads to $C = (ik)^{-1} \int_{0}^{\infty} g(t)e^{ikt} \, dt$, so that Hallén's equation is

$$\int_{-\infty}^{\infty} K(z - z')I(z')\,dz' = r(z), \quad z \in \mathbb{R}, \tag{8.4}$$

where

$$r(z) = \frac{1}{2ik}e^{ikz}\int_{-\infty}^{z} g(t)\,e^{-ikt}\,dt + \frac{1}{2ik}e^{-ikz}\int_{z}^{+\infty} g(t)\,e^{ikt}\,dt . \tag{8.5}$$

The sole purpose of the present section is to show [1] that

$$r(z) \sim \frac{V}{2\zeta_0}e^{ik|z|} \quad (z \to \pm\infty) \quad (\Im k \geq 0) . \tag{8.6}$$

Equation (8.6) shows that $r(z)$ is exponentially small when $\Im k > 0$. For $\Im k = 0$, comparison with Eq. (5.51) or Eq. (11) of [2] shows that the right-hand side of Eq. (8.6)—which is independent of the frill parameters a and b—coincides with the right-hand side for the case where the antenna is driven by a delta-function generator. This is a logical result: far from the feed, the vector potential is independent of the feed details. Equation (8.6) is used repeatedly in [1], whose primary purpose is to explore the consequences of the nonsolvability discussed in Section 3.6.1.

8.1.2   PRELIMINARIES

We base our derivation on some preliminaries. First, for large t and $k > 0$, $g(t)e^{-ikt}$ is nonoscillatory, with

$$g(t)\,e^{-ikt} = \frac{ikV}{2\zeta_0 \ln(b/a)}\left[\frac{ik\,(a^2 - b^2)}{2}\frac{1}{t^2} + O\left(\frac{1}{t^3}\right)\right] \quad (t \to +\infty) . \tag{8.7}$$

Equation (8.7), which also holds for $\Im k > 0$, can be shown by elementary means from Eq. (8.2) (Problem 8.1).

Our second set of preliminaries deals with certain limiting values of the two integrals

$$I_{\pm}\left(\zeta, \xi, z_1, z_2\right) = \int_{z_1}^{z_2} \frac{\exp\left[i\zeta\left(\sqrt{t^2 + \xi^2} \pm t\right)\right]}{\sqrt{t^2 + \xi^2}}\,dt, \quad \xi, z_1, z_2 \in \mathbb{R}, \quad \xi \neq 0 , \tag{8.8}$$

which can be evaluated with a change of variable $u = \sqrt{t^2 + \xi^2} \pm t$. The result is

$$I_{\pm}\left(\zeta, \xi, z_1, z_2\right) = \pm E_1\left[-i\zeta\left(\sqrt{z_1^2 + \xi^2} \pm z_1\right)\right] \mp E_1\left[-i\zeta\left(\sqrt{z_2^2 + \xi^2} \pm z_2\right)\right] , \tag{8.9}$$

where we used the definition Eq. (A.3) of the exponential integral E_1. When $\zeta \neq 0$ with $\Im\zeta \geq 0$, the integrals $I_+\left(\zeta, \xi, z, \infty\right)$ and $I_-\left(\zeta, \xi, -\infty, z\right)$ are seen from (8.8) to be convergent; from (8.9) and the asymptotic approximation $E_1(z) \sim e^{-z}/z \ (z \to \infty)$ (see Section A.2.2 of Appendix A), we obtain

$$I_+ (\zeta, \xi, z, \infty) = \int_z^\infty \frac{\exp\left[i\zeta\left(\sqrt{t^2 + \xi^2} + t\right)\right]}{\sqrt{t^2 + \xi^2}} \, dt = E_1\left[-i\zeta\left(\sqrt{z^2 + \xi^2} + z\right)\right] \qquad (8.10)$$

and

$$I_- (\zeta, \xi, -\infty, z) = \int_{-\infty}^z \frac{\exp\left[i\zeta\left(\sqrt{t^2 + \xi^2} - t\right)\right]}{\sqrt{t^2 + \xi^2}} \, dt = E_1\left[-i\zeta\left(\sqrt{z^2 + \xi^2} - z\right)\right]. \qquad (8.11)$$

We now proceed with our derivation.

8.1.3 DERIVATION OF EQ. 8.6

Equations (8.2) and Eq. (8.5) show that $r(z) = r(-z)$, so it suffices to show Eq. (8.6) for positive z. We write Eq. (8.5) as

$$r(z) = \frac{1}{2ik}e^{ikz}\left[\int_{-\infty}^\infty g(t)\,e^{-ikt}\,dt - \int_z^{+\infty} g(t)e^{-ikt}\,dt + e^{-i2kz}\int_z^{+\infty} g(t)e^{ikt}\,dt\right]. \qquad (8.12)$$

Both for $\Im k > 0$ and $\Im k = 0$, it is a consequence of Eq. (8.7) that the first two integrals in Eq. (8.12) are convergent for all finite z, and that the second one is $o(1)$ as $z \to +\infty$. From Eqs. (8.2) and (8.10), the third integral in Eq. (8.12) is

$$\int_z^{+\infty} g(t)\,e^{ikt}\,dt = \frac{ikV}{2\zeta_0 \ln(b/a)}\left[I_+ (k, a, z, \infty) - I_+ (k, b, z, \infty)\right]$$

$$\qquad\qquad (8.13)$$

$$= \frac{ikV}{2\zeta_0 \ln(b/a)}\left\{E_1\left[-ik\left(\sqrt{z^2 + a^2} + z\right)\right] - E_1\left[-ik\left(\sqrt{z^2 + b^2} + z\right)\right]\right\}.$$

By Eq. (8.13) and $E_1(z) \sim e^{-z}/z$, it follows that the third term inside the brackets in Eq. (8.12) (i.e., $e^{-i2kz}\int_z^{+\infty} g(t)\,e^{ikt}\,dt$) is also $o(1)$ as $z \to +\infty$. Therefore, the asymptotic behavior of $r(z)$ is provided by the $O(1)$ (first) term in Eq. (8.12),

$$r(z) \sim \frac{e^{ikz}}{2ik}\int_{-\infty}^\infty g(t)\,e^{-ikt}\,dt \quad (z \to +\infty). \qquad (8.14)$$

It remains to substitute Eq. (8.2) into Eq. (8.14) and calculate the integral. With the aid of Eq. (8.11), we obtain

$$\int_{-\infty}^{\infty} g(t) \, e^{-ikt} dt = \frac{ikV}{2\zeta_0 \ln(b/a)} \lim_{z \to +\infty} \left[I_- \left(k, a, -\infty, z \right) - I_- \left(k, b, -\infty, z \right) \right]$$

$$= \frac{ikV}{2\zeta_0 \ln(b/a)} \lim_{z \to +\infty} \left\{ E_1 \left[-ik \left(\sqrt{z^2 + a^2} - z \right) \right] - E_1 \left[-ik \left(\sqrt{z^2 + b^2} - z \right) \right] \right\} ,$$

(8.15)

where the limit is needed because the integral $I_- \left(k, a, -\infty, +\infty \right)$ does not converge. Carrying out the limiting process in Eq. (8.15), we obtain (Problem 8.2)

$$\int_{-\infty}^{\infty} g(t) \, e^{-ikt} dt = \frac{ikV}{\zeta_0} .$$

(8.16)

Equations (8.14) and (8.16) yield the desired Eq. (8.6) for $z \to +\infty$. An alternative to Eq. (8.15) method of deriving Eq. (8.16) is discussed in Problem 8.3.

8.2   OSCILLATIONS WITH THE APPROXIMATE KERNEL: CASE OF DELTA-FUNCTION GENERATOR

8.2.1   INTEGRAL EQUATION: NONSOLVABILITY

By [2, 3, 4], Eq. (5.51), or the discussions in Section 8.1.1, the integral equation for the current $I(z)$ on an infinite antenna driven by a delta-function generator is

$$\int_{-\infty}^{\infty} K \left(z - z' \right) I \left(z' \right) dz' = \frac{1}{2\zeta_0} V e^{ik|z|}, \quad z \in \mathbb{R} ,$$

(8.17)

where, as before, we let $K(z)$ stand for the approximate kernel given in Eq. (3.51). For reasons explained in detail in [3] and summarized in Section 3.6.1, the corresponding integral equation for a *finite* antenna of length $2h$ is nonsolvable. In this section,

1. We show that Eq. (8.17) (for the infinite antenna) is also nonsolvable.

2. We apply a "numerical method" to Eq. (8.17); this method has an infinite number of basis functions, each of finite length z_0.

3. For the case of nonzero z_0, we find a closed-form expression for the solution provided by the numerical method.

4. For the case of small z_0, finally, we develop an asymptotic formula for the solution.

In 4., the procedure we follow includes the rudiments of Laplace's method (Chapter 4) and makes use of the Poisson summation formula (PSF, Chapter 6). The asymptotic formula finally developed is useful because, as shown in [2] via numerical results, it also pertains to the numerical solution for the realistic case of the *finite* antenna. This asymptotic formula demonstrates the consequences of nonsolvability and helps distinguish them from the important (but separate) effects of roundoff error and matrix ill-conditioning.

Let $\overline{K}(\zeta)$ stand for the Fourier transform

$$\overline{K}(\zeta) = \int_{-\infty}^{\infty} K(z)e^{i\zeta z}dz .$$ (8.18)

For ζ real, this was found in Eq. (5.44) to be

$$\overline{K}(\zeta) = \begin{cases} \frac{i}{4}H_0^{(1)}\left(a\sqrt{k^2 - \zeta^2}\right), & |\zeta| < k , \\ \frac{1}{2\pi}K_0\left(a\sqrt{\zeta^2 - k^2}\right), & |\zeta| > k . \end{cases}$$ (8.19)

The two expressions in Eq. (8.19) are analytic continuations of one another [3]. By Eq. (A.49) (or see the detailed discussion in Section 5.4.1), $\overline{K}(\zeta)$ is exponentially small when $|\zeta|$ is large, a property that we will use repeatedly.

Generally, integral equations that have the form of Eq. (8.17) can be solved by exploiting the convolution property of the Fourier transform: for $\Im k > 0$ (so that the surrounding medium is slightly lossy), (8.17) implies

$$\overline{I}(\zeta)\,\overline{K}(\zeta) = \frac{ikV}{\zeta_0}\frac{1}{k^2 - \zeta^2} ,$$ (8.20)

where $\overline{I}(\zeta)$ is the Fourier transform of the solution. Equation (8.20) is like Eq. (5.52), with the approximate kernel in place of the exact one. Since $\overline{K}(\zeta)$ is exponentially small, Eq. (8.20) would imply—even for a lossless medium—that $\overline{I}(\zeta)$ is exponentially large. Because no function (satisfying mild admissibility conditions) can have an exponentially large Fourier transform, we have a contradiction verifying that Eq. (8.17) cannot have a solution. We note that, by the discussions in Section 5.4.2, the Fourier transform of the exact kernel is not exponentially small; in that case, one can find the closed-form solution Eq. (5.53).

8.2.2 NUMERICAL METHOD: SOLUTION FOR NONZERO DISCRETIZATION LENGTH

In this section, we apply Galerkin's method with pulse functions to (8.17). Specifically, we seek an approximate solution in the form

$$I(z) \cong \sum_{n=-\infty}^{\infty} I_n u_n(z), \quad z \in \mathbb{R} ,$$ (8.21)

where $u_n(z)$ are the pulse functions given by

$$u_n(z) = \begin{cases} 1, & (n - 1/2)z_0 < z < (n + 1/2)z_0 , \\ 0, & \text{otherwise} . \end{cases} \tag{8.22}$$

Note that we are dividing the entire real axis into segments of length z_0. Assuming initially that $\Im k > 0$, we substitute Eq. (8.21) into (8.17), multiply by $u_n(z)$, and integrate from $z = -\infty$ to $z = +\infty$ to obtain

$$\sum_{n=-\infty}^{\infty} A_{l,n} I_n = B_l \quad \text{or} \quad \sum_{n=-\infty}^{\infty} A_{l-n} I_n = B_l, l \in \mathbb{Z} . \tag{8.23}$$

In Eq. (8.23), the matrix elements $A_{l,n} = A_{l-n}$ are double integrals; our use of the notation A_{l-n} instead of $A_{l,n}$ is due to the easily shown fact that the elements depend on $|l - n|$ only and not on l and n separately. We can further show (Problem 8.4) that the double integral can be reduced to a single integral:

$$A_l = A_{-l} = \int_0^{z_0} (z_0 - z) \left[K(z + l z_0) + K(z - l z_0) \right] dz, \quad l \in \mathbb{Z} . \tag{8.24}$$

The B_l in Eq. (8.23) are

$$B_l = \frac{V}{2\zeta_0} \int_{(l-\frac{1}{2})z_0}^{(l+\frac{1}{2})z_0} \exp\left(ik\,|z|\right) dz = \begin{cases} \dfrac{i2V}{\zeta_0 k}\sin^2\left(\dfrac{kz_0}{4}\right) + \dfrac{V}{\zeta_0 k}\sin\left(\dfrac{kz_0}{2}\right), & \text{if } l = 0, \\[3mm] \dfrac{V}{\zeta_0 k}\sin\left(\dfrac{kz_0}{2}\right) e^{i|l|kz_0}, & \text{if } l = \pm 1, \pm 2, \dots . \end{cases} \tag{8.25}$$

The solution of the doubly infinite Toeplitz system Eq. (8.23) can be found in closed form in the following manner. Multiply each side by $e^{il\theta}$, where $-\pi < \theta \le \pi$, and sum with respect to l. Interchange the order of summation, and introduce the Fourier series

$$\overline{A}(\theta) = \sum_{l=-\infty}^{\infty} A_l e^{il\theta}, \overline{B}(\theta) = \sum_{l=-\infty}^{\infty} B_l e^{il\theta}, \overline{I}(\theta) = \sum_{l=-\infty}^{\infty} I_l e^{il\theta} . \tag{8.26}$$

One obtains $\overline{A}(\theta)\overline{I}(\theta) = \overline{B}(\theta)$ so that

$$I_n = \frac{1}{2\pi} \int_{-\pi}^{\pi} \frac{\overline{B}(\theta)}{\overline{A}(\theta)} e^{-in\theta} d\theta \quad (\Im k > 0) . \tag{8.27}$$

Because $\Im k > 0$, the sum for $\overline{B}(\theta)$ converges and with Eq. (8.25), it is easily found that

$$\overline{B}(\theta) = -\frac{iV}{\zeta_0}\frac{2}{k}\sin^2\left(\frac{kz_0}{4}\right) \frac{\cos\left(\dfrac{kz_0}{2}\right) + \cos^2\left(\dfrac{\theta}{2}\right)}{\sin\left(\dfrac{\theta + kz_0}{2}\right)\sin\left(\dfrac{\theta - kz_0}{2}\right)} . \tag{8.28}$$

Substitution of Eq. (8.24) into (8.26) and application of the PSF (6.1) lead to

$$\overline{A}\left(\theta\right) = \sum_{m=-\infty}^{\infty} \int_{-\infty}^{\infty} \int_{0}^{z_0} \left(z_0 - z\right) \left[K\left(z - xz_0\right) + K\left(z + xz_0\right)\right] dz \, e^{i(\theta - 2\pi m)x} \, dx \,, \qquad (8.29)$$

from which it is seen that

$$\overline{A}\left(\theta\right) = z_0 \sum_{m=-\infty}^{\infty} \overline{K}\left(\frac{2\pi m - \theta}{z_0}\right) \frac{\sin^2\left(\theta/2\right)}{\left(\pi m - \theta/2\right)^2} \,, \qquad (8.30)$$

where $\overline{K}$—the Fourier transform of the kernel—is given by Eq. (8.19).

The expression Eq. (8.28) is also meaningful for real k except for $k = \pm\theta/z_0$. Thus, we can analytically continue Eq. (8.27) to real k if the path of integration in Eq. (8.27) passes below the point $\theta = kz_0$ and above the point $\theta = -kz_0$. The final exact expression for the coefficients I_n can therefore be written as

$$I_n = \frac{1}{\pi} \int_0^\pi \frac{\overline{B}\left(\theta\right)}{\overline{A}\left(\theta\right)} \cos\left(n\theta\right) d\theta, \quad n \in \mathbb{Z}, \quad k > 0 \,. \qquad (8.31)$$

8.2.3 ASYMPTOTIC APPROXIMATION FOR SMALL DISCRETIZATION LENGTH

If one replaces the $\overline{K}\left(\zeta\right)$ in (8.30) by the Fourier transform of the exact kernel, then Eq. (8.31) also holds for the exact kernel [2]. For that case, it is shown in [2] that the limit of Eq. (8.31) as $z_0 \to 0$ is precisely the exact solution (5.53) of the integral equation. With the approximate kernel, on the other hand, Eq. (8.31) has no limit as $z_0 \to 0$, just as expected from nonsolvability: the underlying reason is the exponential smallness of the Fourier transform of the kernel,

$$\overline{K}\left(\zeta\right) \sim \frac{1}{2}\sqrt{\frac{1}{2\pi a \left|\zeta\right|}} e^{-a|\zeta|}, \quad \left(a\left|\zeta\right| \to +\infty\right). \qquad (8.32)$$

Equation (8.32) is a consequence of Eq. (8.19) and Eq. (A.49). To look into the nature of the divergence more carefully, we investigate the asymptotic behavior of Eq. (8.31) subject to the conditions

$$\frac{z_0}{a} \to 0 \quad \text{and} \quad n\frac{z_0}{a} = O(1) \,. \qquad (8.33)$$

We observe from Eqs. (8.30), (8.19), and (8.32) that

$$\overline{A}\left(\theta\right) \sim 4z_0\sin^2\left(\frac{\theta}{2}\right) \left[\overline{K}\left(\frac{\theta}{z_0}\right) \frac{1}{\theta^2} + \overline{K}\left(\frac{2\pi - \theta}{z_0}\right) \frac{1}{(2\pi - \theta)^2}\right], \qquad (8.34)$$

as $z_0/a \to 0$, uniformly for $0 \le \theta \le \pi$. The RHS of (8.34) consists of the terms $m = 0$ and $m = 1$ in the summation (8.30), with all other terms neglected. We keep two terms rather than one because the two terms become equal when $\theta = \pi$ (we encountered a similar situation in Section 6.1.3). Substituting Eq. (8.34) into Eq. (8.31) and setting $\varphi = \pi - \theta$ give us

$$I_n \sim \frac{(-1)^n}{4\pi z_0} \int_0^{\pi} \frac{\overline{B}(\pi - \varphi)/\cos^2(\varphi/2)}{\overline{K}\left(\frac{\pi - \varphi}{z_0}\right)/(\pi - \varphi)^2 + \overline{K}\left(\frac{\pi + \varphi}{z_0}\right)/(\pi + \varphi)^2} \cos(n\varphi)\, d\varphi , \qquad (8.35)$$

where the path of integration passes above the point $\varphi = \pi - kz_0$. Because of (8.32) and Eq. (8.33), the main contribution to the integral in Eq. (8.35) comes from a narrow region near $\varphi = 0$ (or $\theta = \pi$—this stresses the necessity of keeping two terms in Eq. (8.34)). We can therefore neglect the contribution $\int_1^{\pi}$ and replace the upper limit π in Eq. (8.35) by 1 (any other choice of order 1 will do just as well). Then, we can apply Eq. (8.32) for both $\overline{K}\left((\pi - \varphi)/z_0\right)$ and $\overline{K}\left((\pi + \varphi)/z_0\right)$. This was not possible in the original interval, in which the argument of $\overline{K}\left((\pi - \varphi)/z_0\right)$ became small at $\varphi = \pi$. The aforementioned approximations, and the change of variable $x = \varphi a/z_0$ in the resulting integral lead to

$$I_n \sim \frac{1}{\sqrt{2\pi}} kz_0 \sqrt{\frac{z_0}{a}} (-1)^n e^{\pi \frac{a}{z_0}} f\left(\frac{z_0}{a}, ka, n\frac{z_0}{a}\right) , \qquad (8.36)$$

where f is the integral

$$f\left(\frac{z_0}{a}, ka, n\frac{z_0}{a}\right) = \int_0^{\frac{a}{z_0}} g\left(x; \frac{z_0}{a}, ka\right) \cos\left(n\frac{z_0}{a}x\right) dx , \qquad (8.37)$$

in which

$$g\left(x; \frac{z_0}{a}, ka\right) = \frac{\overline{B}\left(\pi - \frac{z_0}{a}x\right) / \left(kz_0^2 \cos^2\left(\frac{z_0}{2a}x\right)\right)}{e^x\left(\pi - \frac{z_0}{a}x\right)^{-5/2} + e^{-x}\left(\pi + \frac{z_0}{a}x\right)^{-5/2}} . \qquad (8.38)$$

The function f depends on its argument z_0/a because a/z_0 is the upper limit of integration in Eq. (8.37), and also because g depends on z_0/a. It can be verified from Eq. (8.28) that (apart from the factor V/ζ_0), g is indeed a function of x, z_0/a, and ka.

   We now approximate f by the first two terms in its Maclaurin expansion (about the point $z_0/a = 0$), keeping ka and nz_0/a fixed. The required differentiations can be carried out using Leibniz's theorem for differentiation of integrals, which is [5]

$$\frac{d}{dx} \int_{\alpha(x)}^{\beta(x)} h(x, y)\, dy = h(x, \beta(x))\beta'(x) - h(x, \alpha(x))\alpha'(x) + \int_{\alpha(x)}^{\beta(x)} \frac{\partial h}{\partial x}\, dy . \qquad (8.39)$$

Thus, the desired expansion can be found by expanding g in powers of z_0/a, and integrating from 0 to ∞ in Eq. (8.37). With Eq. (8.28),

$$
g\left(x; \frac{z_0}{a}, ka\right) = -i\frac{V}{\zeta_0}\frac{\pi^{5/2}}{16}\frac{1}{\cosh x}\left(1 - \frac{5}{2\pi}\frac{z_0}{a}x\tanh x\right)
$$

$$
+ O\left(\left(\frac{z_0}{a}\right)^2\right) \quad \left(\frac{z_0}{a}\to 0\right) . \tag{8.40}
$$

When Eq. (8.40) is substituted into (8.37) and the upper limit of integration is set to ∞, two integrals occur. Both can be found in the usual tables of integrals, or evaluated by symbolic programs. Substitution of the resulting formula for $f(z_0/a, ka, nz_0/a)$ into Eq. (8.36) yields the final result

$$
I_n \sim -i\frac{V}{\zeta_0}\frac{\pi^3}{32\sqrt{2}}kz_0\sqrt{\frac{z_0}{a}}(-1)^n\exp\left(\frac{a\pi}{z_0}\right)\frac{1}{\cosh\left(\frac{\pi}{2}\frac{z_0}{a}n\right)}
$$

$$
\times\left[1 - \frac{5}{2\pi}\frac{z_0}{a} + \frac{5}{4}n\left(\frac{z_0}{a}\right)^2\tanh\left(\frac{\pi}{2}\frac{z_0}{a}n\right)\right] , \tag{8.41}
$$

where the quantity in brackets is simply a correction factor. Equation (8.41) clearly reveals the consequences of nonsolvability: when the pulse width is small, the numerical method yields an exponentially large, purely imaginary "driving-point admittance" and a large, purely imaginary, rapidly oscillating "current," at least for points on the antenna not too far from the driving point.

On the other hand, with nz_0 fixed, using properties of Bessel functions given in Appendix A, it is possible to show (Problem 8.5) that,

$$
\lim_{z_0\to\infty}\text{Re}\left\{\frac{I_n}{V}\right\} = \frac{4k}{\pi\zeta_0}\int_0^k\frac{J_0\left(a\sqrt{k^2-\zeta^2}\right)\cos\left(\zeta nz_0\right)}{(k^2-\zeta^2)\left[J_0^2\left(a\sqrt{k^2-\zeta^2}\right) + Y_0^2\left(a\sqrt{k^2-\zeta^2}\right)\right]}d\zeta , \tag{8.42}
$$

so that the numerical method yields a finite real part of the current. It is seen from Eq. (8.134) of [3] that this real part is very close to the corresponding quantity when the exact kernel is used and that the two real parts become identical in the limit $ka\to 0$.

Since Eq. (8.41) was derived only for the infinite antenna, many numerical checks were performed in [2] before concluding that (8.41) also gives good numerical results for the finite antenna. For example, it was checked that the closeness improved as z_0/a decreases. Furthermore, it was verified that the correction factor in Eq. (8.41) indeed improves upon the approximation. When carrying out the numerical checks, it was important to distinguish from effects due to roundoff: as evidenced by the matrix condition numbers in Fig. 3 of [1], the latter effects are also very important. The relevance of Eq. (8.42) to the finite antenna is that we should not expect oscillations near the center of the antenna in the real part; this was also brought out by numerical calculations in [2].

8.3   ON THE NEAR FIELD DUE TO OSCILLATING CURRENT

8.3.1   STATEMENT OF PROBLEM

The oscillating currents I_n arising from the numerical solution of Hallén's equation for the antenna of infinite length were discussed in Section 8.2. The purpose of [6] is to examine the near field generated by the I_n at a small distance ρ and to bring out similarities to the well-known phenomena of superdirectivity and surface waves. The investigation of [6] is shown to also pertain to the finite antenna and is relevant to a method [4] of post-processing the oscillating numerical solutions so as to obtain a useful, smooth current distribution.

If, in addition to Eq. (8.33), the conditions

$$\frac{\rho}{a} \ll 1 \quad \text{and} \quad \frac{\rho}{z_0} = O(1) \tag{8.43}$$

are satisfied, then it is shown in [6] that the variation of the near magnetic field is given by the function $\eta(\rho/z_0)$, where

$$\eta(x) = \frac{8x}{\pi} \sum_{m=1}^{\infty} \frac{1}{2m-1} K_1((2m-1)\pi x), \quad x \geq 0, \tag{8.44}$$

where K_1 is the modified Bessel function. When x is large, an excellent approximation to Eq. (8.44) results by keeping the $m = 1$ (dominant) term in Eq. (8.44),

$$\eta(x) \sim \frac{8x}{\pi} K_1(\pi x) \quad (x \to +\infty). \tag{8.45}$$

Equation (8.45) can be further simplified by replacing K_1 by the dominant term from Eq. (A.49). However, the interest is in moderate (i.e., not too large) values of $x = \rho/z_0$ (see Eq. (8.43)) for which Eq. (8.45) turns out to be well-suited [6].

The purpose of the present section, which is taken from Appendix A of [6], is to show the less obvious small-x approximation

$$\eta(x) \sim 1 - 2x + (2\ln 2)x^2 \quad (x \to 0^+), \tag{8.46}$$

which consists of the first three terms in the Maclaurin expansion of $\eta(x)$. In Problem 8.6, the reader is asked to show that the numerical accuracy achieved by the two simple expressions (8.45) and Eq. (8.46) is very high. In the course of showing Eq. (8.46), we will derive the integral representation

$$\eta(x) = 4x^2 \int_1^{\infty} \frac{\sqrt{t^2-1}}{\sinh(\pi x t)} \, dt, \quad x \geq 0. \tag{8.47}$$

In the derivation that follows, we use properties of Bessel and other functions that can be found in [5] and [7].

8.3.2 DERIVATION OF EQ. 8.46

Using the integral representation $K_1(z) = z \int_1^\infty e^{-zt} \sqrt{t^2 - 1} \, dt$ (z real) [5] we can write Eq. (8.44) as

$$\eta(x) = 8x^2 \sum_{m=1}^\infty \int_1^\infty e^{-(2m-1)\pi xt} \sqrt{t^2 - 1} \, dt \, . \tag{8.48}$$

Moving the sum inside the integral and then summing the resulting geometric series, we get

$$\eta(x) = 4x^2 \int_1^\infty \frac{\sqrt{t^2 - 1}}{\sinh(\pi xt)} \, dt = \frac{4}{\pi^2} \int_{\pi x}^\infty \frac{\sqrt{u^2 - (\pi x)^2}}{\sinh u} \, du \, , \tag{8.49}$$

where the second expression follows from the first by setting $u = \pi xt$. We have thus shown Eq. (8.47). Since $K_1(x) \sim x^{-1}$ as $x \to 0^+$ [5], Eq. (8.44) gives

$$\eta(0) = \frac{8}{\pi^2} \sum_{m=1}^\infty \frac{1}{(2m-1)^2} = 1 \, , \tag{8.50}$$

where we used a tabulated sum [7]. Equation (8.50) can also be obtained from the second expression Eq. (8.49), as $\int_0^\infty u \operatorname{cschu} du = \pi^2 / 4$ [7].

Differentiating Eq. (8.44) and using the recurrence relation $K_1'(z) = -K_0(z) - K_1(z)/z$ [5], we obtain an expression for the derivative of $\eta(x)$,

$$\eta'(x) = -8x \sum_{m=1}^\infty K_0((2m-1)\pi x) \, . \tag{8.51}$$

In Eq. (8.51), use the integral representation $K_0(z) = \int_1^\infty e^{-zt} / \sqrt{t^2 - 1} \, dt$ ($z \in \mathbb{R}$) [5] and apply the procedure that led to Eq. (8.49) to obtain

$$\eta'(x) = -4x \int_1^\infty \frac{dt}{\sinh(\pi xt) \sqrt{t^2 - 1}} = -4x \int_{\pi x}^\infty \frac{du}{\sinh u \sqrt{u^2 - (\pi x)^2}} \, . \tag{8.52}$$

Alternatively, we can obtain the second expression in Eq. (8.52) from the second expression in Eq. (8.52) using Leibniz's theorem Eq. (8.39) for differentiation of an integral. As a consequence of the first expression Eq. (8.52),

$$\eta'(0) = -\frac{4}{\pi} \int_1^\infty \frac{dt}{t\sqrt{t^2 - 1}} = -2 \, , \tag{8.53}$$

where the integral was evaluated by setting $\sin u = 1/t$.

Differentiating the first expression in (8.52) and then setting $u = \pi x t$ yields

$$\eta''(x) = -4 \int_1^\infty \frac{\sinh(\pi x t) - \pi x t \cosh(\pi x t)}{\sinh^2(\pi x t)\sqrt{t^2 - 1}}\, dt = -4 \int_{\pi x}^\infty \frac{\sinh u - u \cosh u}{\sinh^2 u \sqrt{u^2 - \pi^2 x^2}}\, du. \quad (8.54)$$

Setting $x = 0$ in the second form Eq. (8.54) yields

$$\eta''(0) = -4 \int_0^\infty \frac{\sinh u - u \cosh u}{u \sinh^2 u}\, du. \quad (8.55)$$

Note that the integral in Eq. (8.55) is convergent. Equation (8.55) can be rearranged to give

$$\eta''(0) = -4 \int_0^\infty u^{-1}\left(\frac{1}{\sinh u} - \frac{1}{u}\right) du - 4\lim_{\varepsilon \to 0}\left(\int_\varepsilon^\infty \frac{du}{u^2} - \int_\varepsilon^\infty \frac{\cosh u}{\sinh^2 u}\, du\right). \quad (8.56)$$

If we set $\sinh u = t$ in the third integral in Eq. (8.56), we see that the third integral can be rewritten with an identical integrand $(1/u^2)$ as the second integral; we can then combine those two integrals. The first integral in Eq. (8.56) is a special (limiting) case of the tabulated Mellin transform $\int_0^\infty u^{z-1}(\operatorname{csch} u - u^{-1})\, du = 2(1 - 2^{-z})\Gamma(z)\zeta(z)$ [7], where $\zeta(z)$ is Riemann's zeta function. It follows that

$$\eta''(0) = -4\lim_{z \to 0}\left[2(1 - 2^{-z})\Gamma(z)\zeta(z)\right] - 4\lim_{\varepsilon \to 0}\left(\int_\varepsilon^{\sinh \varepsilon} \frac{du}{u^2}\right). \quad (8.57)$$

By explicit calculation and use of $\sinh \varepsilon = \varepsilon + O(\varepsilon^3)$ as $\varepsilon \to 0$, we see that the second limit in Eq. (8.57) is zero. The first limit can be found by substituting $\Gamma(z) = \Gamma(z+1)/z$, applying l'Hospital's rule, and using $\zeta(0) = -1/2$ and $\Gamma(1) = 1$. The result is

$$\eta''(0) = 4\ln 2. \quad (8.58)$$

Equation (8.46), finally, consists of the first three terms in the Maclaurin expansion of $\eta(x)$, with $\eta(0)$, $\eta'(0)$, and $\eta''(0)$ found from Eq. (8.50), (8.53), and Eq. (8.58).

8.4   SUPPLEMENTARY REMARKS

The elementary idea of using Fourier series to solve doubly infinite Toeplitz systems (Section 8.2.2) has numerous generalizations and applications; we refer the interested reader to [8], the standalone chapter in [9], and the section on inverse systems and deconvolution in [10].

In Section 8.2 we saw that, with the approximate kernel and the delta-function generator, the main consequence of nonsolvability in Hallén's equation is the appearance of rapid oscillations near the driving point. When the antenna is finite, oscillations also occur near the endpoints [2]. Oscillations near the endpoints are also present when the antenna is driven by the magnetic frill generator [1]. In that case, however, there are no oscillations near the driving point; this can be explained by the solvability of the integral equation for the *infinite* antenna [1] (as we saw in Section 3.6.1, the integral equation for the finite antenna is nonsolvable). The Method of Auxiliary Source (MAS) (briefly encountered in Section 3.6.2) is also associated with oscillations [11, 12], as are related methods [13]. Refs. [14, 15] discuss relations between oscillations and superdirectivity for the case of MAS.

8.5   PROBLEMS

8.1.   Show Eq. (8.7).

8.2.   Show Eq. (8.16) by carrying out the limiting process in (8.15) with the aid of Eq. (A.9).

8.3.   With the aid of Eq. (8.18) and Eq. (8.19), show from Eq. (8.2) that

$$\int_{-\infty}^{\infty} g(t)e^{-ikt}dt = \frac{ikV}{\zeta_0 \ln(b/a)} \lim_{\zeta \to k} \left[K_0\left(a\sqrt{\zeta^2 - k^2}\right) - K_0\left(b\sqrt{\zeta^2 - k^2}\right) \right] . \quad (8.59)$$

Then, carry out the limiting process in (8.59) and arrive at Eq. (8.16).

8.4.   Verify Eq. (8.24).

8.5.   Verify Eq. (8.42).

8.6.   Equation (8.46) is a small-x approximation to (8.44) and Eq. (8.45) is a large-x approximation. Numerically demonstrate that (i) Eq. (8.46) always overestimates Eq. (8.44); (ii) Eq. (8.45) always underestimates Eq. (8.44); (iii) the transition point (where the errors in the two formulas become equal in absolute value) is $x = 0.368$; (iv) if, for a given x, one uses the better of the two formulas, the maximum error is only 1.7%.

REFERENCES

[1] G. Fikioris, J. Lionas, and C. G. Lioutas, "The use of the frill generator in thin-wire integral equations," *IEEE Trans. Antennas Propagat.*, vol. 51, no. 8, pp. 1847–1854, August 2003. DOI: 10.1109/TAP.2003.815412. 144, 151, 155

[2] G. Fikioris and T. T. Wu, "On the application of numerical methods to Hallen's equation," *IEEE Trans. Antennas Propagat.*, vol. 49, no. 3, pp. 383–392, March 2001. DOI: 10.1109/8.918612. 144, 146, 147, 149, 151, 155

[3] T. T. Wu, "Introduction to linear antennas," ch. 8 in *Antenna Theory*, pt. I, R. E. Collin and F. J. Zucker, Eds. New York: McGraw-Hill, 1969. 146, 147, 151

[4] G. Fikioris, P. J. Papakanellos, Th. K. Mavrogordatos, N. Lafkas, and D. Koulikas, "Eliminating unphysical oscillations arising in Galerkin solutions to classical integral equations of antenna theory," *SIAM J. Appl. Math.*, vol. 71, no. 2, pp. 559–585, 2011. DOI: 10.1137/100785727. 146, 152

[5] F. W. J. Olver, D. W. Lozier, R. F. Boisvert, and C. W. Clark, *Digital Library of Mathematical Functions*, National Institute of Standards and Technology from `http://dlmf.nist.gov/`http://dlmf.nist.gov/, §1.5.22, §10.30.2, §10.32.8. 150, 152, 153

[6] G. Fikioris, P. J. Papakanellos and Th. K. Mavrogordatos, "Surface-wave and superdirectivity aspects of effective current for linear antennas," accepted for publication in *SIAM J. Appl. Math.* 152

[7] A. P. Prudnikov, Y. A. Brychkov, and O. I. Marichev, *Integrals and Series, Vol. 1*, Gordon & Breach, Amsterdam, The Netherlands, 1998, Entries 2.4.5.9, 2.4.9.7, 5.1.41. 152, 153, 154

[8] H. Widom, "Toeplitz Matrices," in I. I. Hirschman, Jr., Ed., *Studies in Real and Complex Analysis*, Math. Assoc. of America, Englewood Cliffs, NJ: Prentice-Hall, 1965. 154

[9] B. M. McCoy and T. T. Wu, *The two-dimensional Ising model*, Cambridge, MA: Harvard University Press, 1973, Chapter 8. 154

[10] J. G. Proakis and D. G. Manolakis, *Digital Signal Processing, 3rd Ed.* Upper Saddle River, NJ: Prentice Hall, 1996, §4.6. 154

[11] G. Fikioris, "On two types of convergence in the Method of Auxiliary Sources," *IEEE Trans. Antennas Propagat.*, vol. 54, no. 7, pp. 2022–2033, July 2006. DOI: 10.1109/TAP.2006.877171. 155

[12] G. Fikioris and I. Psarros, "On the phenomenon of oscillations in the Method of Auxiliary Sources," *IEEE Transactions Antennas Propagat*, vol. 55, no. 5, pp. 1293–1304, May 2007. DOI: 10.1109/TAP.2007.895621. 155

[13] G. Fikioris, N. L. Tsitsas, and G. K. Charitos, "Spurious oscillations in a combined Method-of-Auxiliary-Sources/Extended-Integral-Equation solution to a a simple scattering problem," *Journal of Quantitative Spectroscopy and Radiative Transfer*, vol. 123, July 2013. DOI: 10.1016/j.jqsrt.2013.01.004. 155

[14] P. Andrianesis and G. Fikioris, "Superdirective-type near fields in the Method of Auxiliary Sources (MAS)," *IEEE Trans. Antennas Propagat.*, vol. 60, no. 6, pp. 3056–3060, June 2012. DOI: 10.1109/TAP.2012.2194671. 155

[15] S. P. Skobelev, 'Comments (on [14], with authors' reply)," *IEEE Trans. Antennas Propagat.*, vol. 61, no. 4, p. 2360, April 2013. 155

APPENDIX A

Special Functions

In this appendix, we give definitions and some properties—mostly asymptotic approximations or expansions—for the special functions encountered in this book. We additionally describe some important reference works for special functions.

Some properties are given without proof. But most derivations are either provided in full or outlined, with the reader asked to fill in details. This appendix is an important part of this book for several reasons: (i) most properties developed here are used elsewhere in the book, so familiarity with them is necessary; (ii) conversely, many of the asymptotics in this appendix are excellent examples of the concepts and techniques developed elsewhere in this book; (iii) the properties and special functions herein—these are often used in applications—are better comprehended if one works out derivations on one's own.

A.1 PRELIMINARIES

There seems to be no generally accepted definition of a special function, or of a special function of mathematical physics. Temme [1] calls "a function 'special' when the function, just like the logarithm, the exponential and trigonometric functions (the elementary transcendental functions), belongs to the toolbox of the applied mathematician, the physicist or engineer." For Borwein and Crandall [2], "special functions are nonelementary functions about which a significant literature has developed because of their importance in either mathematical theory or in practice." They add, "We certainly prefer that this literature include the existence of excellent algorithms for their computation." Today, it is very important that a number of computer packages—such as MAPLE and MATHEMATICA—can handle many special functions, both numerically and symbolically.

Special functions, besides being useful in their own right, are related in a natural manner to asymptotics: historically, many general-purpose asymptotic methods grew out of techniques originally applied to special functions. Today, a learner can greatly benefit by practicing asymptotic methods to integral or series representations of special functions, or to the equations they satisfy.

Each special function has numerous properties, many of which are useful for numerical computation and have been incorporated in computer packages. Apart from integral and series representations, asymptotic expansions and approximations, and equations relating the special function to other functions, properties can include inequalities, recurrence relations, polynomial and rational approximations, closed-form expressions for integrals and sums involving the special function, etc. Currently, perhaps the most useful reference work on special functions and their properties is the print and CD-ROM *NIST Handbook of Mathematical Functions* (NHMF) [3] (published in 2010) and the freely available, online version of NHMF, the *Digital Library of Functions* (DLMF) [4]. The two versions of this work were edited by Olver, Lozier, Boisvert, and Clark. The work has been hailed as a "monumental achievement" [5] and "a treasure for the mathematical and scientific communities, one that will be used and valued for decades" [6]. It is a greatly enlarged, authoritative replacement of the well-known *Handbook of Mathematical Functions*, edited by Abramowitz and Stegun and published in 1964 by NIST (then known as the National Bureau of Standards) [7]. In addition to listing properties, NHMF and DLMF serve to standardize definitions and notations for special functions, to illustrate the importance of each special function in applications, to give detailed graphs, and to describe methods of computation. For all equations and other technical information, NHMF and DLMF either provide references to the literature for proofs, or describe steps that can be followed to construct a proof.

Another very useful source is the Wolfram Functions Site [8]. This is a vast, well-organized collection of formulas involving special functions. Formulas are given without proofs.

Among works that systematically include proofs (rather than references to proofs), let us mention the classic books by Whittaker and Watson [9] and Watson [10]; the latter concerns Bessel and Hankel functions, modified Bessel functions, and related functions. A further classic work is the so-called Bateman Project [11]; here, the proofs are less detailed than in [9] and [10]. More recent works that include proofs are the books by Temme [1] and Andrews, Askey, and Roy [12].

Apart from some notation differences in Section A.4.2, the definitions and properties that follow can be found in NHMF and DLMF. Looking up the relevant equations, studying pertinent graphs and related formulas, etc., is a good way to familiarize oneself with [3] or [4]. In what follows, z denotes a complex variable and x a real and positive variable.

A.2   EXPONENTIAL, SINE, AND COSINE INTEGRALS

A.2.1   DEFINITIONS AND SMALL-ARGUMENT EXPANSIONS

The sine, cosine, and exponential integrals $\mathrm{Si}(z)$, $\mathrm{Ci}(z)$, and $E_1(z)$ are defined by

$$\mathrm{Si}(z) = \int_0^z \frac{\sin t}{t} dt \,, \tag{A.1}$$

$$\mathrm{Ci}(z) = -\int_z^{+\infty} \frac{\cos t}{t} dt, \quad z \neq 0 \,, \tag{A.2}$$

and

$$E_1(z) = \int_z^\infty \frac{e^{-t}}{t} dt, \quad z \neq 0 \,. \tag{A.3}$$

In Eqs. (A.2) and (A.3), the paths of integration do not cross the negative real axis or pass through the origin, where the integrand has a pole. As with the closed definition of the logarithm (Section 3.5.4), the two functions thus defined are the principle values of respective multivalued functions, with branch cuts as in Fig. 3.5(a).

The Maclaurin series of $\mathrm{Si}(z)$ can be obtained by expanding the integrand in Eq. (A.1) using (3.23) and integrating term-by-term. The result is

$$\mathrm{Si}(z) = \sum_{n=0}^{\infty} \frac{(-1)^n}{(2n+1)(2n+1)!} z^{2n+1} . \tag{A.4}$$

The radius of convergence is readily verified to be infinite, showing that $\mathrm{Si}(z)$ is an entire function.

Equation (A.4) is an asymptotic power series of $\mathrm{Si}(z)$ as $z \to 0$. In Problems A.2 and A.3, the reader is asked to show that

$$\mathrm{Ci}(z) = \ln z + \gamma + \sum_{n=1}^{\infty} \frac{(-1)^n}{2n(2n)!} z^{2n} \tag{A.5}$$

and

$$E_1(z) = -\ln z - \gamma - \sum_{n=1}^{\infty} \frac{(-1)^n}{n!n} z^n , \tag{A.6}$$

where $\gamma = 0.57721566\ldots$ is Euler's constant, discussed in Chapter 3. Each of the two power series in Eqs. (A.5) and (A.6) has an infinite radius of convergence. Thus, $\mathrm{Ci}(z)$ and $E_1(z)$ are analytic except for logarithmic singularities at the origin; furthermore, the two power series are the Maclaurin series of the entire functions $\mathrm{Ci}(z) - \ln z - \gamma$ and $E_1(z) + \ln z + \gamma$.

Written explicitly, the first few terms of the power series in (A.4)–(A.6) give small-argument asymptotic approximations to the respective functions. Thus,

$$\mathrm{Si}(z) = z - \frac{z^3}{18} + \frac{z^5}{600} + O(z^7) \quad (z \to 0) , \tag{A.7}$$

$$\mathrm{Ci}(z) = \ln z + \gamma - \frac{z^2}{4} + \frac{z^4}{96} + O(z^6) \quad (z \to 0) , \tag{A.8}$$

and

$$E_1(z) = -\ln z - \gamma + z - \frac{z^2}{4} + \frac{z^3}{18} - \frac{z^4}{96} + O(z^5) \quad (z \to 0) . \tag{A.9}$$

A.2.2 LARGE-ARGUMENT EXPANSIONS

Let x be real and positive. The large-argument asymptotic expansion of $E_1(x)$ can be found using Watson's lemma (Problem 4.5), integration by parts (Section 1.3.2 and Chapter 5), or—because $f(x) = e^x E_1(x)$ satisfies (1.22) and (1.23)—via the formal series solution in Section 1.3.1 (a rigorous justification of this last procedure is beyond the scope of this book; the interested reader can consult Chapter 7 of [13] or §2.7 of [4]). These methods all lead to the divergent asymptotic expansion

$$E_1(x) \sim \frac{e^{-x}}{x} \sum_{n=0}^{\infty} \frac{(-1)^n n!}{x^n} \quad (x \to \infty) , \tag{A.10}$$

cf. (1.42). We note (without giving a proof) that (A.10) also holds for complex x with ph $x <$ $3\pi/2$ [3, 4].

We now turn to $\mathrm{Si}(x)$ and $\mathrm{Ci}(x)$. Define the auxiliary functions $f(x)$ and $g(x)$ through

$$\mathrm{Si}(x) = \frac{\pi}{2} - f(x) \cos x - g(x) \sin x \tag{A.11}$$

and

$$\mathrm{Ci}(x) = f(x) \sin x - g(x) \cos x \,. \tag{A.12}$$

We can show (Problem A.4) that $f(x)$ and $g(x)$ have the following divergent asymptotic power series

$$f(x) \sim \frac{1}{x}\left(1 - \frac{2!}{x^2} + \frac{4!}{x^4} - \frac{6!}{x^6} + \cdots\right) \quad (x \to +\infty) \,, \tag{A.13}$$

$$g(x) \sim \frac{1}{x^2}\left(1 - \frac{3!}{x^2} + \frac{5!}{x^4} - \frac{7!}{x^6} + \cdots\right) \quad (x \to +\infty) \,. \tag{A.14}$$

The sine and cosine integrals thus possess compound asymptotic expansions, a concept defined in Section 2.4. Eqs. (A.11)–(A.14) imply, in particular, that $\mathrm{Si}(+\infty) = \pi/2$ and $\mathrm{Ci}(+\infty) = 0$.

A.3   COMPLETE ELLIPTIC INTEGRAL OF THE FIRST KIND

It is customary to use the symbol k for the argument of the complete elliptic integral of the first kind $K(k)$. This function is defined by

$$K(k) = \int_0^{\pi/2} \frac{d\theta}{\sqrt{1 - k^2 \sin^2\theta}} \,, \tag{A.15}$$

where, for the purposes of this book, it is sufficient to assume $0 \le k < 1$. In Eq. (A.15) our notation—which is the same as that of NHMF/DLMF [4]—differs from other widely used notations, notably from that of the Wolfram Functions Site [8] and the computer program MATHEMATICA. An alternative to the Eq. (A.15) expression follows by setting $t = \sin\theta$,

$$K(k) = \int_0^1 \frac{dt}{\sqrt{1 - t^2}\sqrt{1 - k^2 t^2}} \,. \tag{A.16}$$

From (A.16) we can show (Problem A.5) that $K(k)$ possesses the Maclaurin series

$$K(k) = \frac{\pi}{2}\sum_{n=0}^{\infty}\left[\frac{(1/2)_n}{n!}\right]^2 k^{2n} = \frac{\pi}{2}\left(1 + \frac{1}{4}k^2 + \frac{9}{64}k^4 + \cdots\right) \,, \tag{A.17}$$

whose radius of convergence is 1. In (A.17), we use the usual notation for Pochhammer's symbol, which we define in (3.17).

A.4 BESSEL AND HANKEL FUNCTIONS

A.4.1 DEFINITIONS AND SMALL-ARGUMENT ASYMPTOTIC APPROXIMATIONS

The Bessel functions of argument z and order ν can be defined by

$$J_\nu(z) = \left(\frac{z}{2}\right)^\nu \sum_{k=0}^\infty \frac{(-1)^k}{k!\Gamma(\nu+k+1)} \left(\frac{z}{2}\right)^{2k} \tag{A.18}$$

and

$$Y_\nu(z) = \frac{J_\nu(z)\cos(\pi\nu) - J_{-\nu}(z)}{\sin(\pi\nu)}. \tag{A.19}$$

In (A.18), the radius of convergence of the power series is infinite. Therefore, $J_\nu(z)$ is an entire function of z when $\nu \in \mathbb{Z}$. In this case, it is also true (Problem A.6) that

$$J_{-n}(z) = (-1)^n J_n(z), \quad n \in \mathbb{Z}. \tag{A.20}$$

The situation changes when $\nu \in \mathbb{C}\backslash\mathbb{Z}$. Then, $J_\nu(z)$ has a branch point at $z = 0$ due to the power $(z/2)^\nu$ in Eq. (A.18); this power should be understood as the principal value defined as in (3.42), rendering $J_\nu(z)$ one-valued and analytic in the cut plane of Fig. 3.5(a). When $\nu \in \mathbb{C}\backslash\mathbb{Z}$, it follows from Eq. (A.19) that $Y_\nu(z)$ is also one-valued and analytic in the aforementioned cut plane. When $\nu \in \mathbb{Z}$, (A.20) shows that (A.19) becomes indeterminate; in that case, (A.19) is to be understood by its limiting value.

The Hankel functions can be defined by

$$H_\nu^{(1)}(z) = J_\nu(z) + iY_\nu(z) \tag{A.21}$$

and

$$H_\nu^{(2)}(z) = J_\nu(z) - iY_\nu(z). \tag{A.22}$$

The first few terms in (A.18) give small-z asymptotic approximations to $J_\nu(z)$. Thus,

$$J_\nu(z) = \frac{1}{2^\nu\Gamma(\nu+1)} \left[z^\nu - \frac{1}{4(\nu+1)}z^{\nu+2}\right] + O(z^{\nu+4}) \quad (z \to 0), \quad \nu \neq -1, -2, \dots, \tag{A.23}$$

from which it follows that $J_0(0) = 1$ and $J_n(0) = 0$ for $n = 1, 2, \dots$. In (A.23), we excluded $\nu = -1, -2, \dots$ because $1/\Gamma(\nu+1) = 0$ (see Section 3.1). In these excluded cases, an asymptotic approximation to $J_\nu(z)$ follows from (A.20) and (A.23).

Equations (A.18) and (A.19) can be used to obtain a (convergent) small-z asymptotic expansion for $Y_\nu(z)$. The dominant term of this expansion can be found if one distinguishes the cases $\Re\nu > 0$, $\Re\nu = 0$ and $\Re\nu < 0$. Let us only examine the first case: when $\Re\nu > 0$, (A.19) and the first term in (A.23) give $Y_\nu(z) \sim -(2/z)^\nu / [\Gamma(1-\nu)\sin\pi\nu]$ $(z \to 0)$; Eq. (3.7) then gives

$$Y_\nu(z) \sim -\frac{1}{\pi} 2^\nu \Gamma(\nu) \frac{1}{z^\nu} \quad (z \to 0), \quad \Re\nu > 0. \tag{A.24}$$

Equation (A.24) shows, in particular, that $Y_1(z), Y_2(z), \ldots$ are algebraically unbounded for small z.

We now turn to the case of zero order. For $n = 0, 1, 2, \ldots$ we can show (Problem A.8) that

$$
Y_n(z) = (-1)^n Y_{-n}(z) = -\frac{(z/2)^{-n}}{\pi} \sum_{k=0}^{n-1} \frac{(n-k-1)!}{k!} \left(\frac{z}{2}\right)^{2k} + \frac{2}{\pi} \ln\left(\frac{z}{2}\right) J_n(z) - \frac{(z/2)^n}{\pi}
$$

$$
\sum_{k=0}^{\infty} (-1)^k \frac{\psi(k+1) + \psi(n+k+1)}{k!(n+k)!} \left(\frac{z}{2}\right)^{2k}, \quad n = 0, 1, 2, \ldots,
\tag{A.25}
$$

where ψ is the psi function defined in (3.12). We will further explore (A.25) in Section A.4.3. What interests us now is the special case

$$
Y_0(z) = \frac{2}{\pi} \ln\left(\frac{z}{2}\right) J_0(z) - \frac{2}{\pi} \sum_{k=0}^{\infty} \frac{(-1)^k \, \psi(k+1)}{(k!)^2} \left(\frac{z}{2}\right)^{2k}.
\tag{A.26}
$$

Equation (A.26) can be used to obtain the small-argument asymptotic expansion of $Y_0(z)$; the relevant asymptotic sequence (Definition 2.4) is the same as that in (2.16) and involves logarithms. The first few terms in the expansion are

$$
Y_0(z) = \frac{2}{\pi} \ln\frac{z}{2} + \frac{2\gamma}{\pi} - \frac{1}{2\pi} z^2 \ln\frac{z}{2} + z^2 \frac{1-\gamma}{2\pi} + O(z^4 \ln z) \quad (z \to 0).
\tag{A.27}
$$

Thus, $Y_0(z)$ is unbounded for small z (as are $Y_1(z), Y_2(z), \ldots$), but its singularity at the origin is logarithmic rather than algebraic.

Small-z expansions of $H_0^{(1)}(z)$ and $H_0^{(2)}(z)$ follow immediately from (A.21), (A.22), and (A.26); these two functions also have branch points at the origin and are logarithmically singular there.

A.4.2   LARGE-ARGUMENT ASYMPTOTIC EXPANSIONS

Let x be real and positive. We will give the large-x asymptotic expansions of the Bessel and Hankel functions without proofs. Our notation here resembles the one in [1] and [7] and somewhat differs from the notation of [4]. Define the auxiliary functions $P(\nu, x)$ and $Q(\nu, x)$ through

$$
H_\nu^{(1)}(x) = \sqrt{\frac{2}{\pi x}} \left[P(\nu, x) + iQ(\nu, x) \right] e^{i\left(x - \frac{\pi\nu}{2} - \frac{\pi}{4}\right)}
\tag{A.28}
$$

and

$$
H_\nu^{(2)}(x) = \sqrt{\frac{2}{\pi x}} \left[P(\nu, x) - iQ(\nu, x) \right] e^{-i\left(x - \frac{\pi\nu}{2} - \frac{\pi}{4}\right)}.
\tag{A.29}
$$

By (A.21) and (A.22), these are tantamount to

$$
J_\nu(x) = \sqrt{\frac{2}{\pi x}} \left[P(\nu, x) \cos\left(x - \frac{\pi\nu}{2} - \frac{\pi}{4}\right) - Q(\nu, x) \sin\left(x - \frac{\pi\nu}{2} - \frac{\pi}{4}\right) \right]
\tag{A.30}
$$

and

$$Y_\nu(x) = \sqrt{\frac{2}{\pi x}} \left[P(\nu, x) \, \sin\left(x - \frac{\pi\nu}{2} - \frac{\pi}{4}\right) + Q(\nu, x) \, \cos\left(x - \frac{\pi\nu}{2} - \frac{\pi}{4}\right) \right]. \quad \text{(A.31)}$$

The auxiliary functions possess asymptotic power series as follows [1, 7]:

$$P(\nu, x) \sim \sum_{k=0}^{\infty} (-1)^k \frac{(\nu, 2k)}{(2x)^{2k}} = 1 - \frac{(4\nu^2 - 1)(4\nu^2 - 9)}{128} \frac{1}{x^2} + \cdots \quad (x \to +\infty), \quad \text{(A.32)}$$

$$Q(\nu, x) \sim \sum_{k=0}^{\infty} (-1)^k \frac{(\nu, 2k+1)}{(2x)^{2k+1}} = \frac{4\nu^2 - 1}{8} \frac{1}{x} + \cdots \quad (x \to +\infty), \quad \text{(A.33)}$$

where $(\nu, k) = \Gamma(1/2 + \nu + k)/\left[k!\Gamma(1/2 + \nu - k)\right]$ $(k = 0, 1, 2, \ldots)$ is Hankel's symbol. It follows from (A.28) and (A.29) that the Hankel functions possess asymptotic expansions of the Poincaré type, whose first few terms are

$$H_\nu^{(1)}(x) = \sqrt{\frac{2}{\pi x}} \left[1 + i \frac{4\nu^2 - 1}{8} \frac{1}{x} - \frac{(4\nu^2 - 1)(4\nu^2 - 9)}{128} \frac{1}{x^2} + O\left(\frac{1}{x^3}\right) \right]$$

$$e^{i\left(x - \frac{\pi\nu}{2} - \frac{\pi}{4}\right)}, \quad (x \to +\infty) \quad \text{(A.34)}$$

and

$$H_\nu^{(2)}(x) = \sqrt{\frac{2}{\pi x}} \left[1 - i \frac{4\nu^2 - 1}{8} \frac{1}{x} - \frac{(4\nu^2 - 1)(4\nu^2 - 9)}{128} \frac{1}{x^2} + O\left(\frac{1}{x^3}\right) \right]$$

$$e^{-i\left(x - \frac{\pi\nu}{2} - \frac{\pi}{4}\right)}, \quad (x \to +\infty). \quad \text{(A.35)}$$

On the other hand, $J_\nu(x)$ possesses a compound asymptotic expansions (as defined in Section 2.4) consisting of the equality (A.30) and the asymptotic expansions (A.32) and (A.33). Similarly, the compound asymptotic expansion of $Y_\nu(x)$ consists of (A.31)–(A.33). It is usual—but, as discussed in Section 2.3.2, not rigorous—to describe the "leading terms" by expressions such as

$$J_\nu(x) \sim \sqrt{\frac{2}{\pi x}} \left[\cos\left(x - \frac{\pi\nu}{2} - \frac{\pi}{4}\right) - \frac{4\nu^2 - 1}{8x} \sin\left(x - \frac{\pi\nu}{2} - \frac{\pi}{4}\right) \right] \quad (x \to +\infty) \quad \text{(A.36)}$$

and

$$Y_\nu(x) \sim \sqrt{\frac{2}{\pi x}} \left[\sin\left(x - \frac{\pi\nu}{2} - \frac{\pi}{4}\right) + \frac{4\nu^2 - 1}{8x} \cos\left(x - \frac{\pi\nu}{2} - \frac{\pi}{4}\right) \right] \quad (x \to +\infty). \quad \text{(A.37)}$$

A.4.3   LARGE-ORDER ASYMPTOTIC APPROXIMATIONS

Different asymptotic approximations apply when the order, ν, is large. As $\nu \to +\infty$, the first term in (A.18) is easily seen to be much larger than all other terms. Therefore,

$$J_\nu(z) \sim \frac{1}{\Gamma(\nu+1)}\left(\frac{z}{2}\right)^\nu \quad (\nu \to +\infty, \; z \neq 0) . \tag{A.38}$$

A similar formula for $Y_\nu(z)$ can be shown (Problem A.9) from (A.18), (A.19), and (A.25). It is

$$Y_\nu(z) \sim -\frac{1}{\pi}\Gamma(\nu)\left(\frac{z}{2}\right)^{-\nu} \quad (\nu \to +\infty, \; z \neq 0) . \tag{A.39}$$

Alternatives to (A.38) and (A.39) result by applying Stirling's formula (Section 3.1). They are

$$J_\nu(z) \sim \frac{1}{\sqrt{2\pi\nu}}\left(\frac{ez}{2\nu}\right)^\nu \quad (\nu \to +\infty, \quad z \neq 0) , \tag{A.40}$$

$$Y_\nu(z) \sim -\sqrt{\frac{2}{\pi\nu}}\left(\frac{ez}{2\nu}\right)^{-\nu} \quad (\nu \to +\infty, \quad z \neq 0) . \tag{A.41}$$

From the above formulas, it follows that $J_\nu(z)$ approaches zero extremely rapidly as $\nu \to +\infty$, while $|Y_\nu(z)|)$ increases extremely rapidly. Combining (A.40) and (A.41) with (A.21) and (A.22) we obtain

$$H_\nu^{(1)}(z) \sim -\frac{i}{\pi}\Gamma(\nu)\left(\frac{z}{2}\right)^{-\nu} , H_\nu^{(2)} \sim \frac{i}{\pi}\Gamma(\nu)\left(\frac{z}{2}\right)^{-\nu} \quad (\nu \to +\infty, \quad z \neq 0) , \tag{A.42}$$

or, alternatively,

$$H_\nu^{(1)}(z) \sim -i\sqrt{\frac{2}{\pi\nu}}\left(\frac{ez}{2\nu}\right)^{-\nu} , \quad H_\nu^{(2)}(z) \sim +i\sqrt{\frac{2}{\pi\nu}}\left(\frac{ez}{2\nu}\right)^{-\nu} \quad (\nu \to +\infty, \quad z \neq 0) . \tag{A.43}$$

A.4.4   ADDITION THEOREM FOR HANKEL FUNCTION OF ORDER ZERO

The so-called addition theorem is [11]

$$H_0^{(1)}\left(\sqrt{x_1^2 + x_2^2 - 2x_1 x_2 \cos\theta}\right)$$

$$= \sum_{n=-\infty}^{\infty} J_n\left(\min\{x_1, x_2\}\right) H_n^{(1)}\left(\max\{x_1, x_2\}\right) e^{in\theta} , \tag{A.44}$$

where $x_1 > 0$, $x_2 > 0$, and θ is real. In the right-hand side of (A.44), the quantity multiplying $e^{in\theta}$ is the nth Fourier-series coefficient for the periodic function of θ on the left-hand side. If one forms a triangle by including an angle θ between two sides x_1 and x_2, the argument of the Hankel function on the left-hand side of (A.44) equals the triangle's third side.

A.5 MODIFIED BESSEL FUNCTIONS

Here, we give relations similar to those of the previous section for the modified Bessel functions, which are defined by

$$I_\nu(z) = \left(\frac{z}{2}\right)^\nu \sum_{k=0}^\infty \frac{(z/2)^{2k}}{k!\,\Gamma(\nu+k+1)} \tag{A.45}$$

and

$$K_\nu(z) = \frac{\pi}{2} \frac{I_{-\nu}(z) - I_\nu(z)}{\sin(\pi\nu)} . \tag{A.46}$$

For reasons obvious from (A.18) and (A.19), $I_\nu(z)$ and $K_\nu(z)$ are also called Bessel functions of imaginary argument. Let us note that our notation for $K_\nu(z)$, which is the same as that of [4], differs from the one in Whittaker and Watson [9].

As previously, (A.45) is the asymptotic expansion of $I_\nu(z)$ for small z. Relations like (A.24) can easily be found for $K_\nu(z)$, while for the case of zero order we have [4]

$$K_0(z) = -\ln\frac{z}{2} - \gamma + O(z^2 \ln z) \quad (z \to 0). \tag{A.47}$$

For large, positive arguments, $I_\nu(x)$ increases exponentially while $K_\nu(x)$ decreases exponentially according to [4]

$$I_\nu(x) = \frac{e^x}{\sqrt{2\pi x}} \left[1 + \frac{1 - 4\nu^2}{8x} + O\left(\frac{1}{x^2}\right) \right] \quad (x \to +\infty) \tag{A.48}$$

and

$$K_\nu(x) = \sqrt{\frac{\pi}{2x}}\, e^{-x} \left[1 - \frac{1 - 4\nu^2}{8x} + O\left(\frac{1}{x^2}\right) \right] \quad (x \to +\infty) . \tag{A.49}$$

Finally, large order asymptotic approximations are [4]

$$I_\nu(z) \sim \frac{1}{\sqrt{2\pi\nu}} \left(\frac{ez}{2\nu}\right)^\nu \quad (\nu \to +\infty, \quad z \neq 0) \tag{A.50}$$

and

$$K_\nu(z) \sim \sqrt{\frac{\pi}{2\nu}} \left(\frac{ez}{2\nu}\right)^{-\nu} \quad (\nu \to +\infty, \quad z \neq 0) . \tag{A.51}$$

A.6 GENERALIZED HYPERGEOMETRIC FUNCTIONS

The generalized hypergeometric series of order (p, q) is defined as a formal power series in z and is denoted by $_pF_q\,(\alpha_1, \alpha_2, \ldots, \alpha_p;\ \beta_1, \beta_2, \ldots, \beta_q;\ z)$. The expressions for the power-series coefficients involve the p numbers α_k and the q numbers β_k $(p, q = 0, 1, 2, \ldots)$, called upper and lower parameters, respectively. The definition is

$$_pF_q\,(\alpha_1, \ldots, \alpha_p;\ \beta_1, \ldots, \beta_q;\ z) = \sum_{k=0}^\infty \frac{(\alpha_1)_k (\alpha_2)_k \cdots (\alpha_p)_k}{(\beta_1)_k (\beta_2)_k \cdots (\beta_q)_k} \frac{z^k}{k!} , \tag{A.52}$$

where we use the usual notation for Pochhammer's symbol (Section 3.1). In (A.52), all lower parameters are usually assumed to be different from $0, -1, -2, \ldots$. Application of the ratio test for power series and Stirling's formula yields the following cases (see [4] and Section 7.2.3 of [14]):

Case 1: When $p \le q$, the radius of convergence is infinite. In this case, the series in (A.52) defines an entire function, the so-called generalized hypergeometric function.

Case 2: When $p = q + 1$, the radius of convergence is 1. In this case, the generalized hypergeometric function—denoted, again, by $_pF_q(\alpha_1, \ldots, \alpha_p; \beta_1, \ldots, \beta_q; z)$—is defined (i) by the series (A.52) when $|z| < 1$, and (ii) by the analytic continuation of this series when $|z| \ge 1$. The function thus defined is often (but not always, see Problem A.11) multivalued, with a branch point at $z = 1$; in those cases, the symbol $_pF_q(\alpha_1, \ldots, \alpha_p; \beta_1, \ldots, \beta_q; z)$ denotes the principal branch, as defined in Section 7.2.3 of [14].

Our Case 3 is a condition for divergence:

Case 3: When $p \ge q + 2$, the series in (A.52) generally has a zero radius of convergence; for exceptions, see [4] and Section 7.2.3 of [14].

Generalized hypergeometric functions are much more general than the more usual "special functions of mathematical physics." Because of the many parameters involved, many of the usual special functions are in fact special cases of $_pF_q$, see Problem A.11 for examples.

The special case with $p = 2$ and $q = 1$ is the hypergeometric function, encountered in Problem 3.9:

$$_2F_1(a, b; c; z) = F(a, b; c; z) = \sum_{k=0}^{\infty} \frac{(a)_k (b)_k}{(c)_k} \frac{z^k}{k!}. \tag{A.53}$$

Chapter 7 of [14] and the Wolfram Functions Site [8] contain extensive tables which can be searched in a systematic manner to see whether a given $_pF_q$ can be reduced to a more usual special function.

There are many asymptotic approximations and expansions for $_pF_q(\alpha_1, \ldots, \alpha_p; \beta_1, \ldots, \beta_q; z)$ pertaining, for example, to large z, large parameters, or combinations; see [3], the references therein, and [8].

A.7   PROBLEMS

A.1.   Verify the statements made in this appendix concerning radii of convergence of the various power series.

A.2.   Show (A.6) as follows:

(i) Rearranging (A.3), show that

$$E_1(z) = \int_0^1 \frac{e^{-t} - 1}{t}\, dt + \text{Ein}(z) + \int_1^\infty \frac{e^{-t}}{t}\, dt - \ln z,$$

where

$$\text{Ein}(z) = \int\limits_0^z \frac{1 - e^{-t}}{t}\, dt\,. \tag{A.54}$$

(ii) Arrive at (A.6) by determining the Maclaurin series of $\text{Ein}(z)$ and introducing γ through (3.71).

(The notation $\text{Ein}(z)$ for the function in (A.54) is standard [4]. This function is entire. It is sometimes called the complementary exponential integral.)

A.3. Show (A.5) with the aid of (A.6). (See Problem A.2 for a derivation of (A.6).)

A.4. Show (A.13) and (A.14) as follows:

(i) Use contour-integration techniques to show that

$$\text{Si}(+\infty) = \int\limits_0^\infty \frac{\sin t}{t}\, dt = \frac{\pi}{2}\,.$$

(ii) Derive the following integral representations,

$$f(x) = \int\limits_0^\infty \frac{\sin t}{t + x}\, dt; \quad g(x) = \int\limits_0^\infty \frac{\cos t}{t + x}\, dt\,. \tag{A.55}$$

(iii) Use (A.55) and the technique of integration by parts (Chapter 5) to arrive at (A.13) and (A.14).

(iv) Verify that the formal series in (A.13) and (A.14) diverge.

A.5. Derive (A.17) using (A.16) and the binomial expansion. *Hint:* Use the integral defining the so-called beta function [4],

$$B(a, b) = \int\limits_0^1 t^{a-1}(1 - t)^{b-1} dt, \quad \Re a > 0, \quad \Re b > 0\,, \tag{A.56}$$

as well as the following relation [4] between the gamma and beta functions

$$B(a, b) = \frac{\Gamma(a)\Gamma(b)}{\Gamma(a + b)}, \quad \Re a > 0, \quad \Re b > 0\,. \tag{A.57}$$

(For a proof of (A.57), see §12.41 of [9].)

A.6. Show (A.20) from (A.18).

A.7. Obtain a formula similar to (A.24) for the case $\Re\nu < 0$. (Beware of the special cases $\nu = -1/2, -3/2, \ldots$.)

A.8. (i) Show from (A.19) and (A.20) that

$$Y_n(z) = \frac{1}{\pi}\frac{\partial J_\nu(z)}{\partial \nu}\bigg|_{\nu=n} + \frac{(-1)^n}{\pi}\frac{\partial J_\nu(z)}{\partial \nu}\bigg|_{\nu=-n}, \qquad n \in \mathbb{Z}. \qquad (A.58)$$

(ii) Show from (A.18) that

$$\frac{\partial J_\nu(z)}{\partial \nu} = J_\nu(z)\ln\left(\frac{z}{2}\right) - \left(\frac{z}{2}\right)^\nu \sum_{k=0}^{\infty} \frac{(-1)^k\,\psi(\nu+k+1)}{k!\,\Gamma(\nu+k+1)}\left(\frac{z}{2}\right)^{2k}, \qquad (A.59)$$

where ψ is the psi function defined in (3.12).

(iii) Combine (A.58) and (A.59) and, with the aid of (3.72), obtain (A.25).

A.9. (i) Substitute (A.18) into (A.19) to form an expression for $Y_\nu(z)$ involving two power series. In each series, retain the largest term as $\nu \to +\infty$ to show that

$$Y_\nu(z) \sim \frac{\cot\pi\nu}{\Gamma(\nu+1)}\left(\frac{z}{2}\right)^\nu - \frac{1}{\pi}\Gamma(\nu)\left(\frac{z}{2}\right)^{-\nu}. \qquad (A.60)$$

(ii) In the case where ν is a large positive integer, the first term in (A.60) becomes infinite. Except for this case, the second term is dominant and (A.39) holds. Complete the derivation of (A.39) by showing from (A.25) that (A.39) holds even when ν is a large positive integer.

A.10. (i) Show that

$$J_{-\nu}(z) \sim \frac{\Gamma(\nu)(z/2)^{-\nu}}{\pi}\sin\pi\nu + \frac{(z/2)^\nu}{\Gamma(\nu+1)}\cos\pi\nu \qquad (\nu \to +\infty). \qquad (A.61)$$

(ii) Explain why the asymptotic approximation (A.61) should be understood as a compound one. (Compound asymptotic approximations are discussed in Section 2.3.)

(iii) Let z be real and fixed and let ν be the variable, with ν large and positive. Use (A.61) to find approximate values for the zeros of $J_{-\nu}(z)$ (often called ν-zeros) and to roughly sketch $J_{-\nu}(z)$ as a function of ν. Compare with a computer-generated graph of $J_{-\nu}(z)$.

A.11. Show that

$$\frac{d}{dz}\,_pF_q\left(\alpha_1,\ldots,\alpha_p;\ \beta_1,\ldots,\beta_q;\ z\right)$$
$$= \frac{\alpha_1\alpha_2\cdots\alpha_p}{\beta_1\beta_2\cdots\beta_q}\,_pF_q\left(\alpha_1+1,\ldots,\alpha_p+1;\ \beta_1+1,\ldots,\beta_q+1;\ z\right)\ ,$$

$$_2F_1\left(a,b;\ b;\ z\right) = {}_1F_0\left(a;\ z\right) = (1-z)^{-a}\ ,$$

$$\ln z = (z-1)\,_2F_1\left(1,1;2;1-z\right)\ ,$$

$$\mathrm{Si}(z) = z\,_1F_2\left(\frac{1}{2};\frac{3}{2},\frac{3}{2};-\frac{z^2}{4}\right)\ ,$$

$$\sin z = z\,_0F_1\left(\frac{3}{2};-\frac{z^2}{4}\right)\ ,$$

$$K(k) = \frac{\pi}{2}\,_2F_1\left(\frac{1}{2},\frac{1}{2};1;k^2\right)\ ,$$

$$J_\nu(z) = \frac{1}{\Gamma(\nu+1)}\left(\frac{z}{2}\right)^\nu\,_0F_1\left(\nu+1;-z^2/4\right)\ .$$

REFERENCES

[1] N. M. Temme, *Special Functions; an Introduction to the Classical Functions of Mathematical Physics*. New York: Wiley, 1996. DOI: 10.1002/9781118032572. 159, 160, 164, 165

[2] J. M. Borwein and R. E. Crandall, "Closed Forms: What They Are and Why We Care," *Notices of the AMS*, vol. 60, no. 1, pp. 50–65, January 2013. DOI: 10.1090/noti936. 159

[3] F. W. J. Olver, D. W. Lozier, R. F. Boisvert, and C. W. Clark, *Handbook of Mathematical Functions*. Cambridge, UK: Cambridge University Press, 2010. 159, 160, 162, 168

[4] F. W. J. Olver, D. W. Lozier, R. F. Boisvert, and C. W. Clark, *Digital Library of Mathematical Functions*, National Institute of Standards and Technology from `http://dlmf.nist.gov/`. 160, 161, 162, 164, 167, 168, 169

[5] P. J. Davis, "Formulas Galore," Book Review in *SIAM News*, vol. 43, no. 7, Sept. 2010. 160

[6] R. Beals, "Handbooks of Mathematical Functions, Versions 1.0 and 2.0."Book Review in *Notices of the AMS*, vol. 58, no. 8, Sept. 2011, pp. 1115–1118. 160

[7] M. Abramowitz and I. A. Stegun, Eds., *Handbook of Mathematical Functions with Formulas, Graphs, and Mathematical Tables*, (National Bureau of Standards Applied Mathematics Series, Vol. 55.) US Government Printing Office, Washington, D.C., 1964. 160, 164, 165

[8] *Wolfram Functions Site* from `http://functions.wolfram.com/`. 160, 162, 168

[9] E. T. Whittaker and G. N. Watson, *A Course of Modern Analysis, 4th ed.* Cambridge, UK: Cambridge University Press, 1927. (Reprinted, 1997). 160, 167, 169

[10] G. N. Watson, *A Treatise on the Theory of Bessel Functions. 2nd ed.* Cambridge, UK: Cambridge University Press, 1944. (Reprinted, Cambridge Mathematical Library Edition, 1995). 160

[11] Staff of the Bateman Manuscript Project (A. Erdélyi (Editor), W. Magnus, F. Oberhettinger, and F. G. Tricomi (Research Associates)), *Higher Transcendental Functions. Vols. I—III.* New York: McGraw-Hill, 1953. (Reprinted, Malabar, FL: Robert Krieger Publishing Co., 1981.) 160, 166

[12] G. E. Andrews, R. Askey, and R. Roy, *Special Functions.* Cambridge, UK: Cambridge University Press, 1999. DOI: 10.1017/CBO9781107325937. 160

[13] F. W. J. Olver, *Asymptotics and Special Functions.* Natick, MA: A. K. Peters, 1997. 161

[14] A. P. Prudnikov, Yu. A. Brychkov, and O. I. Marichev, *Integrals and Series: More Special Functions,* vol. 3. London: Taylor and Francis, 2002. (Reprint of 1990 ed.) 168

APPENDIX B

On the Convergence/Divergence of Definite Integrals

In this stand-alone appendix, we discuss simple methods for investigating whether a given definite integral converges or diverges. We specifically consider what mathematicians call "improper integrals" (we omit the adjective "improper" throughout this book) in which the integration interval extends to infinity or the integrand is infinite at finite points within the integration interval.

It is possible to have a convergent integral when the integrand is infinite at finite points; but the infinity cannot be too rapid. For an infinite integration interval, it is *not* sufficient for the integrand to vanish at infinity for the integral to converge[1]; the rate of this vanishing is also important. In this appendix, we provide simple, systematic rules of thumb that can help us determine when a given integral converges and apply these rules to a large number of examples. Our integration path is always on the real axis.

B.1 SOME REMARKS ON OUR RULES

Before giving our rules, some clarifications are necessary.

1. In this appendix, and throughout this book, our integrals are "classical" ones. Thus, we consider integrals such as $\int_{-\infty}^{\infty} \exp(ixy)dy$ to be divergent because the double limit

$$\frac{1}{ix}\left(\lim_{M\to\infty} e^{ixM} - \lim_{N\to-\infty} e^{ixN}\right) \tag{B.1}$$

does not exist. A meaning to such integrals can of course be attached with the aid of the Dirac delta function (the above integral is then equal to $2\pi\delta(x)$). Alternatively, one can attach meaning via "Abel summability;" see [1]. In this book, however, we regard the integral as divergent because we choose to use a different mathematical framework.

2. In what follows, we are interested in the behavior of the integrand only at one endpoint of the integration interval, 0 or $+\infty$. Thus, we examine integrals of the type $\int_0^A$ and $\int_A^\infty$ where

---

[1]This is obvious, for example, from Rule 3 later. Conversely, if the integral converges, it is not necessary that the integrand vanishes at infinity. This is seen from the Fresnel integral $\int_0^\infty \cos(\pi t^2/2)dt$, the convergence of which can be shown by changing the variable $x = \pi t^2/2$ and using Rule 5 later.

$A > 0$. In the remaining integration interval, we assume that our integrands are "sufficiently smooth," a property we make no attempt to clarify further.

3. We also assume that the convergence/divergence of the integral can be determined if we replace the integrand by its leading asymptotic approximation. For a "sufficiently small" integration interval—that is, for sufficiently small (large) A for integrals of the form $\int_0^A$ $\left(\int_A^\infty \right)$—such a replacement is usually legitimate.

4. Because we do not precisely specify the above-mentioned limitations on the integrand and the integration interval, the analysis that follows is not rigorous. Despite this, the convergence/divergence rules that follow are very useful in practice. More rigorous treatments can be found in the literature [2, Apendix I], [1]. A historical account of relevant topics can be found in [3].

B.2   RULES FOR DETERMINING CONVERGENCE/DIVERGENCE

We now present our rules, together with justifications.

Rule 1—Algebraic behavior at zero: If

$$g(y) \sim \frac{B}{y^{1-\varepsilon}} \quad \text{as} \quad y \to 0^+ \ (B \neq 0) , \tag{B.2}$$

then the integral $\int_0^A g(y) dy$

(i) converges if $\Re \varepsilon > 0$ and

(ii) diverges if $\Re \varepsilon \leq 0$.

Justification of Rule 1: The rule is valid because

$$\int_0^A \frac{dy}{y^{1-\varepsilon}} = \frac{A^\varepsilon}{\varepsilon} - \lim_{y \to 0^+} \frac{y^\varepsilon}{\varepsilon} \tag{B.3}$$

(here, we assumed $\varepsilon \neq 0$) and the limit in Eq. (B.3) exists and is finite only when $\Re \varepsilon > 0$. For the special case $\varepsilon = 0$, we have

$$\int_0^A \frac{dy}{y} = \ln A - \lim_{y \to 0^+} \ln y , \tag{B.4}$$

in which the limit does not exist, so that the integral diverges.

Rule 2 is a generalization of Rule 1, allowing a logarithmic dependence to multiply the algebraic one.

Rule 2—Algebraic/logarithmic behavior at zero: If δ is real and

$$g(y) \sim B\frac{(-\ln y)^{\delta}}{y^{1-\varepsilon}} \text{ as } y \to 0^+ \ (B \neq 0) \ , \tag{B.5}$$

then the integral $\int_0^A g(y)\, dy$

(i) converges if $\Re\varepsilon > 0$ and

(ii) diverges if $\Re\varepsilon < 0$.

In other words, at least for $\Re\varepsilon \neq 0$, the presence of the logarithm does not affect the convergence or divergence of the integral.

Justification of Rule 2: With a change of variable $-\ln y = t$, we see that

$$\int_0^A \frac{(-\ln y)^{\delta}}{y^{1-\varepsilon}} dy = \int_{-\ln A}^{\infty} e^{-\varepsilon t} t^{\delta} dt \ , \tag{B.6}$$

and it is obvious that the integral in Eq. (B.6) converges if $\Re\varepsilon > 0$ and diverges if $\Re\varepsilon < 0$. (Using the above change of variable, it is also possible to come up with a rule for the case $\Re\varepsilon = 0$, but the said rule is more complicated and depends on δ.)

Rule 3—Algebraic behavior at infinity: If

$$g(y) \sim \frac{B}{y^{1+\varepsilon}} \text{ as } y \to +\infty \ (B \neq 0) \ , \tag{B.7}$$

then the integral $\int_A^{\infty} g(y)\, dy$

(i) converges if $\Re\varepsilon > 0$ and

(ii) diverges if $\Re\varepsilon \leq 0$.

Justification of Rule 3: Similar to that of Rule 1.
The following rule generalizes Rule 3.

Rule 4—Algebraic/logarithmic behavior at infinity: If δ is real and

$$g(y) \sim B\frac{(\ln y)^{\delta}}{y^{1+\varepsilon}} \text{ as } y \to \infty \ (B \neq 0) \ , \tag{B.8}$$

then the integral $\int_0^A g(y)\, dy$

(i) converges if $\Re\varepsilon > 0$ and

(ii) diverges if $\Re\varepsilon < 0$.

Here, as in Rule 2, for $\Re\varepsilon \neq 0$, the presence of the logarithm does not affect the convergence or divergence.

Justification of Rule 4: Similar to that of Rule 2, with a change of variable $\ln y = t$.

Rule 5—Sinusoidal/algebraic behavior at infinity: If x is real with $x \neq 0$ and

$$g(y) = O\left(\frac{1}{y^\varepsilon}\right) \quad \text{as } y \to \infty , \tag{B.9}$$

then the integrals

$$\int_A^\infty g(y)\cos xy\,dy \text{ and } \int_A^\infty g(y)\sin xy\,dy \tag{B.10}$$

(i) converge if $\Re\varepsilon > 0$ and, in particular,

(ii) they converge absolutely if $\Re\varepsilon > 1$.

(Recall that the integral $\int_A^\infty g(y)\,dy$ converges absolutely if $\int_A^\infty |g(y)|\,dy$ converges.) Comparing with Rule 3, we see that the condition on $g(y)$ for convergence is weaker; the convergence for $0 < \Re\varepsilon < 1$ is due to the *oscillations* of the integrand.

Justification of Rule 5: From the inequalities

$$\left|\frac{\sin xy}{y^\varepsilon}\right| \le \frac{1}{y^{\Re\varepsilon}}, \quad \left|\frac{\cos xy}{y^\varepsilon}\right| \le \frac{1}{y^{\Re\varepsilon}} , \tag{B.11}$$

and Rule 3, the absolute convergence (and therefore, the convergence) for $\Re\varepsilon > 1$ follows immediately. For $\Re\varepsilon > 0$, an integration by parts yields

$$\int_A^\infty \frac{\sin xy}{y^\varepsilon}\,dy = -\frac{1}{x}\lim_{y\to\infty}\frac{\cos xy}{y^\varepsilon} + \frac{1}{x}\frac{\cos Ax}{A^\varepsilon} - \frac{\varepsilon}{x}\int_A^\infty \frac{\cos xy}{y^{1+\varepsilon}}\,dy . \tag{B.12}$$

In the right-hand side of Eq. (B.12), the limit exists and is zero, while the last integral converges (absolutely) as we just showed. Thus, for $\Re\varepsilon > 0$, the integral $\int_A^\infty \frac{\sin xy}{y^\varepsilon}\,dy$ converges. The convergence, for $\Re\varepsilon > 0$, of the integral $\int_A^\infty \frac{\cos xy}{y^\varepsilon}\,dy$ can be established in a similar manner.

B.3   EXAMPLES

Applying the above rules, the reader can verify that the following integrals converge:

$$\int_1^\infty \frac{y^2}{\sqrt{y^7+1}}\,dy , \tag{B.13}$$

(the integral $\int_1^\infty \frac{y^3}{\sqrt{y^7+1}}dy$, however, diverges),

$$\int_0^\infty \frac{\sin y}{\sqrt{y}}dy\,, \tag{B.14}$$

$$\int_0^\infty \frac{\sin y}{y}dy\,, \tag{B.15}$$

(but $\int_0^\infty \frac{\cos y}{y}dy$ diverges),

$$\int_1^\infty \sin\left(\frac{1}{y^2}\right)dy\,, \tag{B.16}$$

(the integrand Eq. (B.16) converges absolutely),

$$\int_0^\infty \frac{\cos xy}{\sqrt{1+y^2}}dy,\quad x>0\,, \tag{B.17}$$

$$\int_0^\infty K_0(y)\cos xydy,\quad x>0\,, \tag{B.18}$$

$$\int_0^\infty y^{-1/2}K_0(y)\cos xydy,\quad x>0\,, \tag{B.19}$$

(see Appendix A for the relevant properties of $K_0(y)$)

$$\int_0^\infty \frac{y^{z-1}}{\sqrt{1+y^2}}dy,\quad 0<\Re z<1\,, \tag{B.20}$$

$$\int_0^\infty \frac{y^{z-1}}{e^y-1}dy,\quad \Re z<1\,, \tag{B.21}$$

$$\int_0^\infty \frac{\sin y}{1+y^2}y^{z-1}dy,\quad -1<\Re z<3\,, \tag{B.22}$$

$$\int_0^\infty y^{z-1}\ln\left(y+\sqrt{y^2+1}\right)dy,\quad -1<\Re z<0\,. \tag{B.23}$$

The last four integrals are Mellin transforms, with each restriction on z denoting the corresponding strip of analyticity (SOA), as defined in Chapter 7.

REFERENCES

[1] D. V. Widder, *Advanced Calculus*, 2nd ed. Englewood Cliffs, NJ: Prentice-Hall, 1961 (reprinted, New York: Dover, 1989), Chapt. 10. 173, 174

[2] P. Prudnikov, Yu. A. Brychkov, and O. I. Marichev, *Integrals and Series: More Special Functions*, vol. 3. London, UK: Taylor and Francis, 2002. (Reprint of 1990 ed.). 174

[3] M. A. Evgrafov, "Series and integral representations," in *Analysis, I.* R. V. Gamkrelidze, Ed. New York: Springer, Chapt. I, 1989. (Translated from the Russian by D. Newton). 174

Authors' Biographies

GEORGE FIKIORIS

George Fikioris was born in Boston, MA on December 3, 1962. He received the Diploma of Electrical Engineering from the National Technical University of Athens, Greece (NTUA), in 1986, and the S.M. and Ph.D. degrees in Engineering Sciences from Harvard University in 1987 and 1993, respectively. From 1993 to 1998, he was an electronics engineer with the Air Force Research Laboratory (AFRL), Hanscom AFB, MA. From 1999 to 2002, he was a researcher with the Institute of Communication and Computer Systems at the National Technical University of Athens, Greece (NTUA). In 2002, he joined the faculty of the School of Electrical and Computer Engineering, NTUA, where he is currently an Associate Professor. He is the author or coauthor of 50 papers in technical journals, a list of which can be found in `http://www.ece.ntua.gr/images/publications/george-fikioris-publications.pdf`

Together with R. W. P. King and R. B. Mack, he has co-authored *Cylindrical Antennas and Arrays,* Cambridge University Press, 2002. He is the author of *Mellin-Transform Method for Integral Evaluation: Introduction and Applications to Electromagnetics,* Morgan & Claypool, 2007. His research mostly deals with fundamental and mathematical aspects of electromagnetics and antennas, especially wire antennas and antenna arrays.

Dr. Fikioris is a senior member of the IEEE (Antennas & Propagation, Microwave Theory & Techniques, and Education Societies), and a member of the Technical Chamber of Greece.

IOANNIS TASTSOGLOU

Ioannis Tastsoglou was born in Kavala, Greece, in 1986. He received the Telecommunications and Electronics degree from the Hellenic Air Force Academy, Athens, Greece, in July 2007. He is currently a Lieutenant in the Hellenic Air Force. He is concurrently working toward the Ph.D. degree in the School of Electrical and Computer Engineering of the National Technical University of Athens. His research interests include electromagnetics, antennas, and applied mathematics.

ODYSSEAS N. BAKAS

Odysseas N. Bakas was born in Athens, Greece, in 1989. He received the Diploma in Electrical and Computer Engineering from the National Technical University of Athens (NTUA), Greece, in 2011. He is currently pursuing a Master's Degree at the School of Applied Mathematical and Physical Sciences. His research interests include electromagnetics and applied mathematics.

Index

Abel, N. H., 32

Abel summability, 173

Ablowitz, M. J., 138

Abramowitz, M., 160

analytic continuation, 36, 41-44, 59, 61-65, 147, 149

    and multivalued functions, 44-47

    and nonsolvability, 53-58

    applications to electromagnetics and antennas, 53-58, 60, 65, 75

    of generalized hypergeometric series, 168

    of Mellin transform, 131

analytic (regular) functions, 35, 39, 41-44, 54, 57

    and periodicity, 119

    at infinity, 39

    integrals as, 43-44

    special functions as, 161, 163

Andrews, G. E., 160

antenna arrays

    array factor for, 3

    circular, 110, 113

    cylindrical, 119-122

    linear, 110-115

    pseudopotential, 114-115

approximate kernel, 53-55, 75, 97-98, 143, 146-151, 155

    Fourier transform of, 98, 144, 149

Askey, R., 160

asymptotic approximations, 23-33

    addition of, 25, 26, 30, 32, 33

applications to electromagnetics and antennas, 1-4, 14-18, 76, 84, 97-99, 100, 113, 122, 136-138, 143-155

    compound, 27-30, 32, 33, 79

    definition of, 23, 27

    differentiation of, 26

    for certain Fourier transforms, 96-97

    for pendelum, 4-7, 26

    for special functions, 73-74, 90-91, 132, 159-171. *See also name of individual special function*

    integration of, 26

    multiplication of, 25

    produced by integration by parts, 92-94

    produced by Laplace's method, 69-73, 76

    produced by Poisson Summation Formula, 105, 113

    simple examples of, 1-20, 24-26

    truncated Taylor and Maclaurin series as, 26

asymptotic expansions, 30-31

    and divergent series, 32

    and exponentially smaller terms, 92

    and Watson's lemma, 78-79, 102

    compound, 31, 32, 79, 137, 140, 162, 165

    for solutions to differential equations, 14, 19

    for special functions, 79, 81, 85-86, 90-92, 140, 159-169. *See also name of individual special function*

generalized, 32

history, 31-32

of the Poincaré type, 30, 36, 165. *See also*
    Poincaré asymptotic expansions

optimal truncation of, 13, 91

produced by integration by parts, 91

produced by Mellin-transform method,
    129-140

produced by Poisson Summation
    Formula, 105, 109

simple examples of, 14, 31

asymptotic formula. *See* asymptotic
    approximation

asymptotic power series, 31, 78, 140, 161,
    162, 165

with all terms zero, 95, 98

asymptotic sequence, 30-31, 109, 125

asymptotic series. *See* asymptotic expansions

Bachmann, P., 31

beamwidth, 85

Bender, C. M., 27

Bessel functions ($J_\nu(z)$, $Y_\nu(z)$), 163-166

and nonradiating currents, 121

large-argument asymptotics of, 26,
    28-30, 31, 79, 164-165

large-order asymptotic approximations
    of, 166, 170

Mellin transform of, 130

small-argument asymptotic
    approximations of, 163-164

beta function ($B(a, b)$), 169

binomial coefficient $\left(\dbinom{m}{n}\right)$, 61

binomial expansion, 40, 61

binomial theorem, 61

Bloch phase shift, 115

Boisvert, R. F., 160

Borwein, J. M., 159

branch (of multivalued function), 48-53, 59,
    63-65

branch cut, 49, 52-54, 63, 102, 161

branch point, 44-47, 54, 57, 58, 63, 163, 164,
    168

point at infinity as, 47

Brychkov, Yu. A., 75, 130

Carrier, G. F., xviii

Cauchy, A-L, 32

Cauchy principal value, 82-85

Cauchy's integral formula, 50

circuits (electric), 4, 18, 19

circle of convergence, 39

circular arrays. *See* antenna arrays, circular

circular-loop antenna, 74-77, 99

Clark, C. W., 160

closed form, xvii, 1, 55, 58, 107, 132, 137,
    146, 147, 148, 159

comb function, 105

complementary error function (erfc x), 89-92

asymptotic expansion of, 91

complementary exponential integral
    (Ein(z)), 169

complete elliptic integral of the first kind
    ($K(k)$), 6-7, 103, 162-163

Maclaurin series of, 7, 162

compound asymptotic approximations. *See*
    asymptotic approximations,
    compound

compound asymptotic expansions. *See*
    asymptotic expansions, compound

cosine integral (Ci (z)), 96, 160-162

large-argument expansion of, 162

small-argument expansion of, 161

Crandall, R. E., 159

delta-function generator, 53, 59-60, 74, 77,
    99, 100, 144, 146-151, 155

logarithmic singularity of current near, 100

differentiation under the integral sign, 44

Digital Library of Mathematical Functions (DLMF), xviii, 16, 59, 160

disk of convergence, 39, 41

distributions, 101, 106, 118

divergent series, 19, 27, 32, 91, 118

duplication formula, 36

Einstein's needle radiation, 136

error function (erf x), 91-92

Euler, L., 32, 35, 59, 63

Euler's constant, Euler-Mascheroni constant (γ), 31, 37, 59, 60, 137

Euler numbers (E_n), 62

exact kernel, 97-100, 147, 149, 151
    Fourier transform of, 99, 147
    logarithmic singularity of, 99

exponential integral ($E_1(z)$), 11, 160-162
    large-argument expansion of, 85, 89, 161
    small-argument expansion of, 161

Extended Integral Equation (EIE), 60, 65

Extended Lagrange Inversion Theorem, 80, 85

factorial ($n!$), 36

falling factorial, 59

fast wave, 111-113

Felsen, L. B., xvii

finite-gap source, 74, 77

Fokas, A. S., 138

Fourier transform (cosine, exponential, sine), 28, 89, 92-101, 105-108, 123, 147, 149

free-space impedance (ζ_0), 53, 111, 143

frill generator, 53-54, 74, 77, 143-146, 155

Frobenius, F. G., 19

Fuchs, L., 19

Galileo, 19

Galerkin's method, 147

gamma function ($\Gamma(z)$), 35-37, 59-61, 73-74, 80-81, 131, 169
    and Mellin-transform method, 35, 131-132
    asymptotic expansion of. *See* Stirling's formula
    plot of, 38

Gauss's multiplication formula, 36

geometric series, 40, 41

Geometrical Optics (GO), 4

Geometrical Theory of Diffraction, 4

generalized hypergeometric funtion ($_pF_q$), 132, 167-168, 171

Green's function, 106, 111, 114

Hallén's equation, 53-55, 59-60, 75, 97-100, 143-144, 152, 155
    kernel of. *See* approximate kernel *and* exact kernel
    moment-method solutions of, 100

Hamming, R. W., xvii

Hankel's loop integral, 65

Hankel's symbol ((ν, k)), 165

Hankel functions ($H_\nu^{(1)}(z), H_\nu^{(2)}(z)$), 163-166
    addition theorem for $H_0^{(1)}(z)$, 166
    large-argument asymptotic expansions of, 165
    large-order asymptotic approximations of, 166
    small-argument asymptotic approximations of, 164, 171

highly directive current distributions, 136-138

hill-shaped integrand, 69, 71, 73, 85

Huygens, C., 19

hypergeometric function ($F(a, b; c; z)$, $_2F_1(a, b; c; z)$), 62, 76, 168 impulse train, 105-106

incomplete gamma function ($\Gamma(a, x)$), 86, 91
  asymptotic expansion of, 86

integrals:
  analyticity of, 43-44
  convergence/divergence of, 173-177

integration by parts, 10-14, 83-84, 89-103, 161, 169
  applications to wire antennas, 97-100
  for Fourier transforms, 92-95
  for Laplace transforms, 89-92
  that does not work, 91, 93-94, 97
  that produces boundary terms equal to zero, 94-95

irregular singular point, 19

isochronous, 6, 19

Jordan's lemma, 35, 79, 102

kernel, approximate. *See* approximate kernel

kernel, exact. *See* exact kernel

Kirchhoff's laws, 4

Knopp, K., 59

Lagrange inversion theorem, 80-81, 85

Landau, E., 31

Laplace's method, 69-77, 84-85
  application to circular-loop antenna, 76
  application to wire antennas, 147

Laplace transform, 77, 79, 85, 89-92

Leibniz's theorem, 150, 153

Lozier, D. W., 160

linear antennas. *See* wire antennas

little-Oh (o), 23

logarithm, logarithmic function ($\ln z, \mathrm{Ln}\, z$), 46, 47, 50-53
  closed definition of, 52-53
  Maclaurin series involving, 40

Maclaurin series, 26, 39

and Watson's lemma, 78

application to wire antenas, 150, 152

for complete elliptic integral of first kind, 162

for elementary functions, 16-17, 39-40, 46, 86

manipulations of, 40, 61-62

related to sine, cosine, and exponential integrals, 161, 168-169

magnetic frill generator. *See* frill generator

Marcuvitz, N., xvii

Margetis, D., xvii

Marichev, O. I., 75, 130

Maxwell's equations, 4

Meijer-G function, 132, 135

Mellin-Barnes integrals, 130-132, 135-136
  of first type, 136
  of second type, 136, 137

Mellin convolution, 130-131

Mellin transform, 63, 65, 68, 154, 177. *See also* Mellin-transform method
  and Mellin convolution, 131
  and Mellin-Barnes integrals, 130
  and standard products, 130
  definition of, 129
  elementary properties of, 130
  inversion formula for, 130
  of $J_\nu(x)$, 130
  of $\sin x$ and $\cos x$, 94
  strip of analyticity (SOA) of, 129, 177
  tables of, 130

Mellin-transform method, 35, 76, 129-140
  and steepest-descent contribution, 137
  application to highly directive currents, 136-138
  for asymptotic evaluation of integrals, 129-140
  for exact evaluation of integrals, 129-132
  residue calculations in, 132-133

simple example of, 133-135

Method of Auxiliary Sources (MAS), 53, 55-58, 60, 82-84

and nonsolvability, 55-58

and oscillations, 155

and superdirectivity, 155

critical radius (ρ_{cri}) in, 57, 58, 60, 65

continuous version of, 55

modified Bessel functions ($I_\nu(z)$, $K_\nu(z)$), 167

large-argument asymptotic approximation for $K_0(x)$, 24, 31, 102, 167

large-order asymptotic approximations of, 167

small-argument asymptotic expansion for $K_0(z)$, 31, 167

moment methods, 100

multivalued functions, 15, 16, 20, 44-53, 57, 59, 161, 168

Newton, I., 45, 86

National Institute of Standards and Technology (NIST), 16, 59, 159-160

NIST *Handbook of Mathematical Functions* (NHMF), 159-160

NIST *Digital Library of Mathematical Functions. See Digital Library of Mathematical Functions*

nonradiating currents, 121

nonsolvability, 53-58, 59-60, 75-77, 84, 99, 144, 146-147, 151, 155

order (O), 13, 16-18, 23-27, 30-31, 85, 90, 96-97

Oberhettinger, F., 107

Olver, F. W. J., 31, 160

Orszag, S. A., 27

oscillations:

and asymptotic approximations, 30

associated with Hallén's equation with approximate kernel, 146-155

associated with Method of Auxiliary Sources, 155

of integrands, 93, 95, 176

of pendelum, 4-7, 19

Oseen, C. W., 136

parabolic cylinder function ($U(a, x)$), 85

pendulum, 4-7, 18-19, 26

Physical Theory of Diffraction, 4

Pocklington's equation, 53-55, 59-60, 75, 97-99, 143

kernel of. *See* approximate kernel *and* exact kernel

Pochhammer's symbol ($(z)_n$), 37, 59, 61

Poincaré, H., 32

Poincaré asymptotic expansion, 31. *See also* asymptotic expansion, of Poincaré type

Poisson Summation Formula (PSF), 105-126, 147, 149

and sampling theorem, 107, 125

and trapezoidal rule, 116-119

applications to electromagnetics and antennas, 107-115, 119-122, 147, 149

asymptotic approximation produced by, 105, 113

asymptotic expansion produced by, 105, 109

convergence properties of sum produced by, 107, 109, 113, 123

derivation of, 105-106, 116-117, 126

improper use of, 118

in notation of signal processing, 124

involving functions with continuous periodic extensions, 118-119

for doubly infinite sums, 105-115

for finite sums (PSF-FS), 111, 115-122, 126

for semi-infinite sums, 107, 118, 126

power series. *See* asymptotic power series, Maclaurin series, Taylor series

principal branch, 48-53, 168

principal value:

of functions, 16, 40, 48-53, 57, 63, 65, 163

of integrals. *See* Cauchy principal value

of series, 116, 118

Prudnikov, A. P., 75, 130

psi function ($\psi(z)$), 37, 59, 60, 133

radius of convergence, 10, 19, 39-40, 61-62, 161, 163, 168

rapidly decreasing function, 101

ratio test, 10, 39, 40, 168

recurrence formulas, 10, 159

for gamma function, 36

for psi function, 37

involving $K_1(z)$, 153

reflection formula:

for gamma function, 36, 60

for psi function, 37

removable singularities, 41, 82

residues, 35, 36, 61, 81, 132-133, 139

Riemann, B., 59

Riemann zeta function ($\zeta(z)$), 63, 154

Riemann-Lebesgue lemma, 92-93

Roy, R., 160

saddle-point method, xvii, 85

sampling theorem, 125

Sasiela, R. J., 137

sine integral (Si (z)), 93, 97, 160-162

large-argument expansion, 162

small-argument expansion, 161

slow wave, 110-115, 119-121

Smith chart, 14-18, 20

special functions, xviii, 35-38, 50, 79, 132, 159-172

and generalized hypergeometric functions, 132, 168, 171

and Meijer-G functions, 132

reference works for, 159-160

square root function, 44-49

standard products, 130

standing-wave ratio (SWR), 14-18

stationary-phase method, xvii

steepest-descent method, xvii, 85, 137

Stegun, I. A., 160

Stirling's formula, 36-37, 73-74, 79-81

strip of analyticity (SOA), 129, 177

superdirectivity, 84, 152, 155

surface wave, 112, 152

symbolic programs. *See* symbolic routines

symbolic routines, 6, 40, 81, 84, 107, 108, 130, 139, 151, 159

tables

for simplification of generalized hypergeometric function, 132

for simplification of Meijer-G function, 132

of Fourier transforms, 107

of integrals, 6, 75, 76, 107, 108, 151

of Mellin transforms, 130

Taylor series, 26, 39-40

about infinity, 39, 61

as asymptotic expansions, 30

Temme, N. M., 159, 160

theta functions, 123

time dependence ($e^{j\omega t}$, $e^{-i\omega t}$), xviii, 14, 53, 56, 74, 97, 111, 114, 143

Toeplitz system, 148, 154

transmission line, 14-18, 20

Watson, G. N., 59, 160, 167

Watson's lemma, 77-86, 91, 102, 161

Weierstrass, K., 59, 60

Whittaker, E. T., 59, 160, 167

wire antennas, 1-4, 53-55, 97-100, 143-155

Wolfram Functions Site, 130, 132, 160, 162, 168

Wu, T. T., xviii

Printed in the United States
by Baker & Taylor Publisher Services